A model of the cyclol structure constructed for Dorothy Wrinch by Niels Bohr.

*Even if a new theory should meet an early death, it should not be forgotten; rather, its beauty should be remembered, and history should record our gratitude to it – for bequeathing to us new and perhaps still unexplained experimental facts, and, with them, new problems; and for the services it has thus rendered to the progress of science during its successful but short life.*
— Karl R. Popper

# STRUCTURES OF MATTER AND PATTERNS IN SCIENCE

*inspired by the work and life of Dorothy Wrinch, 1894-1976*

The Proceedings of a Symposium held at
Smith College, Northampton, Massachusetts
September 28-30, 1977
and
Selected papers of Dorothy Wrinch,
from the Sophia Smith Collection

edited by *Marjorie Senechal*
*Department of Mathematics*

Schenkman Publishing Company, Inc.
Cambridge, Massachusetts

Copyright c 1980
Schenkman Publishing Company, Inc.
3 Mt. Auburn Place
Cambridge, MA 02138

**Library of Congress Cataloging in Publication Data**
Main entry under title:

Structures of matter and patterns in science.

    Includes bibliographies
    1. Molecular biology — Congresses. 2. Matter —
Structure — Congresses. 3. Science — Philosophy — Congresses. 4. Wrinch, Dorothy Maude. I. Wrinch, Dorothy
Maud. II. Senechal, Marjorie. III. Smith College.
IV. Sophia Smith Collection.
QH506.S84      501      79-513
ISBN 0-87073-908-5
ISBN 0-87073-909-3 pbk.

*Printed in the United States of America.*

All rights reserved. This book, or parts thereof,
may not be reproduced in any form without
written permission of the publisher.

# FOREWORD

This volume is the record of a symposium entitled "Structures of Matter and Patterns in Science" held in honor of the late Dorothy Wrinch by her colleagues in Smith College. Since Dorothy Wrinch had been associated with Smith College from 1942 to 1971, it was appropriate that the suggestion for a symposium in her honor should have come from the fellow faculty members of the institution of learning where she spent most of her scholarly life.

Dorothy Wrinch, born in Argentina, studied mathematics in Cambridge (England), London, and Oxford. By 1929 she had accumulated a BA, two MA's, one MS, and two DSc's. Judging by the titles of many of her early papers she was especially interested in geometry. Accordingly she continued her training by studying geometry in Vienna in 1931; yet onto the rigorous career in mathematics she grafted an interest in biology, which she also studied in the early 1930's in various laboratories of Europe. Her first paper on proteins was published in 1934 in *Nature*. From this point on, proteins occupied much of her thought and her publication.

At this stage in the history of x-ray crystallography, methods for the determination of the arrangements of atoms in an arbitrary crystal were not yet available, although some crystal structures with only a few atoms per cell could be determined if special conditions prevailed, such as the occurrence of a heavy atom in a special position. The general reason for this was that the periodic pattern of the atoms in a crystal can be represented (and therefore determined) by a Fourier series whose coefficients are the amplitudes of the x-rays scattered by the crystal. These quantities are complex, involving both magnitude and phase: the magnitudes can be measured but unfortunately the phases cannot.

In these circumstances, A.L. Patterson made a direct attempt to find out what kind of information about a crystal could be determined by the use of scattering magnitudes only. In 1934 he was able to demonstrate that a Fourier series whose coefficients were the squares of the scattering magnitudes alone, supplied a function whose high values have origin distances equal to the interatomic vectors in the crystal. When this development came to the attention of Dorothy Wrinch she realized that the Patterson function, as it was called, provided a test for the correctness of any proposed model of a crystallized protein. Accordingly, in late 1938 and early 1939 she published two papers on the geometry of Patterson maps, and in 1940 she discussed the Patterson projection of the structure proposed for the insulin molecule.

Dorothy Wrinch evidently took personal pleasure in applying her geometrical training to studying the patterns of proteins with the aid of that geometrical science, crystallography. Once committed to protein patterns, these provided the direction for her scientific work. She devoted her intellect and energy to this direction with only occasional diversions into crystallographic side lines such as "Twinning and intergrowths" (*Amer. Mineralogist* 32, 1947). Her main energy was channeled into presenting and defending her cyclol hypothesis of protein structure. This was controversial, and now appears to be a lost cause. Yet, considering the limited x-ray criteria available at the time it was proposed, one cannot adversely criticize Dorothy Wrinch for proposing it.

Because of her interest in the crystallography of proteins Dorothy Wrinch customarily attended the meetings of the American Society for X-Ray and Electron Diffraction (the ASXRED), the American Crystallographic Society (the ACS) and their common successor, the American Crystallographic Association (the ACA). Thus she became acquainted with many crystallographers and mineralogists. Needless to say, she was well acquainted with the members of the faculty of Smith College and indeed with the academicians in the general Northampton-Amherst area. In the summers, she joined the biologists at Wood's Hole on Cape Cod, Massachusetts, where she made many friends. Indeed, three of the six who gave formal lectures at the symposium were colleagues and neighbors at Wood's Hole.

The symposium lectures embrace a broad spectrum of science. The first lecture was given by David Harker, a crystallographer who had worked with Dorothy Wrinch in the Chemistry Department of Johns Hopkins University in the early 1940's. His talk, entitled "Colored Lattices", was given at the Sigma Xi luncheon. In this talk,

Harker described some of his recent work on color symmetry, demonstrating how it could be applied to simple symmetrical crystal structures.

On the first evening, Caroline MacGillavry, a crystallographer from the Netherlands, gave a lecture on "Symmetry in Three Kingdoms: Animal, Mineral and Vegetable". Attended by many Smith College students, this talk was made especially interesting by the attention paid to the way the symmetries of living things adapt them to their life styles.

The next morning, Smith faculty members, students and guests met to discuss "Reflections on Symmetry: Truth and Beauty" and "The Specialization Trap".

That afternoon Arthur Loeb, assuming the dual role of geometrist and artist, spoke on "Sculptural Models, Modular Sculptures". His talk was illustrated with models whose geometrical properties he transformed by changes which had a quasi-magical quality.

The second evening Carolyn Cohen gave a lecture entitled "Deciphering Protein Designs". This was an outline of the history and development of our present understanding of the structures of proteins. For the nonchemist, this proved an easy introduction to proteins.

On the morning of the third day, biologist Ruth Hubbard spoke on "Reflections on the Story of the Double Helix". This turned out to be an incredible story which presented an object lesson on how science ought not, and generally does not, advance.

The afternoon of the third day was devoted to discussions by Smith faculty members, students and guests on "Scientific Method?" and "The Wrinch Papers".

The third evening a lecture was given by George Wald on "Life in the Universe". This turned out to be a wide-ranging subject in which the history of the universe, from stars down to protons, neutrons and electrons, and finally chemistry and life, were covered.

The texts of the six lectures, which were very well attended, make up the bulk of this volume.

I would like to close these preliminary remarks with some personal words about Dorothy Wrinch. Before meeting Dorothy I had already had occasion to look up her first three papers on vector maps which discussed the geometry of Patterson functions. These had been published in late 1938 and early 1939. They aided me in formulating my own views of treating this function which led (as treated later in the discussion of "The Wrinch Papers") to a solution of crystal structures through Patterson space. I eventually met Dorothy at one of the Gibson Island meetings sponsored by the National Research Council's Committee on X-ray and Electron Diffraction in the late 1930's and early 1040's. I recall that she gave a talk on Fourier transforms which I only partly understood, so I wrote her a letter later asking clarification of certain points. Her answers to my questions were carefully worded and improved my understanding of these transforms. Eventually when she sought a place to publish an extended paper on Fourier transforms I invited her to make it Monograph No. 2 of the ASXRED. Unfortunately the referee unduly delayed publication of this work, but during the ensuing correspondence, Dorothy and I became good friends.

I remember Dorothy Wrinch as a delightful person, one who knew more about geometry than I, yet willing to discuss with me things of mutual interest. She was a friend to many colleagues, as this record of the symposium attests.

*Martin Buerger*
*June 15, 1978*

# CONTENTS

Foreword   *Martin Buerger* .............................. v

Introduction   *Marjorie Senechal* ........................ 1

I. The Lectures of the Symposium

    Colored Lattices   *David Harker* ..................... 11
    Symmetry in Three Kindgoms   *Caroline MacGillavry* .. 31
    Scuptural Models, Modular Sculptures   *Arthur Loeb* ... 59
    Deciphering Protein Designs   *Carolyn Cohen* ......... 77
    The Story of DNA   *Ruth Hubbard* .................. 117
    Life in the Universe (excerpts)   *George Wald* ........ 139

II. The Dorothy Wrinch Collection

    The Wrinch Papers   *Discussion* .................... 149

    Bibliography ........................................ 171

    Selected Papers from the Wrinch Collection .......... 180

    Chronology of the Life of Dorothy Wrinch .......... 191

Suggested Reading ........................................ 193

Notes .................................................... 197

Acknowledgements ........................................ 201

"The Earth is the most fantastic planet in the solar system. No other planet has children playing, big sisters sitting in the sun."
—from a fourth-grade science report

"What is the relation between those large particles which we call elephants, trees, or men, and these extremely small ones which we call molecules or electrons? Because at the present day the biochemist has often little enough to offer toward the solution of the problem of the origin and maintenance of organic form, the morphologist is apt to suppose that no connections exist, and to acquiesce in an acceptance of the ancient Aristotelian distinction between *materia* and *forma*. This, however, is a counsel of despair."
—Joseph Needham, *Order and Life*, 1936
(dedicated to Dorothy Wrinch
and the other members of the
Theoretical Biology Club)

"Snow crystals, and all the rest besides, have much to teach us about the variety, the beauty, and the very nature of form."
—D'Arcy W. Thompson, *On Growth and Form*

"We are not students of some subject matter but students of problems, and problems may cut right across the borders of any subject matter or discipline."
—Karl Popper,
*Conjectures and Refutations*

"The scientific method is, and must be, disciplined by an orthodoxy which can permit only a limited degree of dissent, and such dissent is fraught with grave risks to the dissenter."
—Michael Polanyi,
"The Potential Theory of Adsorption"

"You and I have much in common in the manner we managed to make our scientific careers less dull than usual."
—Michael Polanyi to Dorothy Wrinch, 1948

"Some think it a matter of course that chance
Should starve good men and bad advance,
That if their neighbors figured plain
As though upon a lighted screen,
No single story would they find
Of an unbroken happy mind,
A finish worthy of the start."

—W.B. Yeats,
"Why Should Not Old Men Be Mad?"

# INTRODUCTION

The discovery in 1912 that a beam of X-rays passed through a crystal produces a diffraction pattern on a photographic plate was one of the most far-reaching of our time. It opened the possibility of determining the atomic structure of both inorganic and organic materials, and led to remarkable advances in scientific knowledge and in technology. But the symmetric pattern produced on the film is an indirect and incomplete clue to the three-dimensional arrangement of the atoms in the crystal through which the beam has passed. Before these possibilities could be realized, the crucial problem of interpreting the beautiful but puzzling diffraction patterns had to be solved.

Thus, the theoretical and experimental science of modern crystallography came into being, and for many years it has attracted to its ranks able scientists — chemists, physicists, mineralogists, mathematicians — who are fascinated by problems in pattern, structure, and form. One of the most remarkable of these was Dorothy Wrinch, an Englishwoman, who made fundamental contributions to x-ray diffraction theory and who was a controversial catalyst in the development of what is now called molecular biology. Her work ranged from philosophy, mathematics, and sociology to physics, chemistry, mineralogy, and biology. "I know of no one," wrote Margery Fry,[1] "who moves more freely in the world of pure thought. Her power of concentration has often surprised me almost as much as her grasp of ideas not only in her own but in cognate subjects . . . She has in many ways the most remarkable intellect I have known in a woman during a lifetime spent largely in the company of academic women in this country."

Dorothy Wrinch spent thirty years in Northampton, Massachusetts, in various research positions in the Department of Physics at Smith College. After her death in 1976 at the age of 82, her voluminous papers were given to the Sophia Smith Collection at the College. The Wrinch Collection is a most valuable resource in the history and sociology of science. To honor her memory and to make the Collection more widely known, a symposium inspired by her work and life was held at Smith College in September 1977. Entitled *Structures of Matter and Patterns in Science*, it explored in one event the variety of interrelated themes her work and life suggest.

The structure of matter, the understanding of life, is one of the grand themes in science. In our time, it has centered on the search for the relation, in Joseph Needham's words, between "those large particles which we call elephants, trees or men, and those extremely small ones which we call molecules or electrons."[2] This search is carried out by human beings, men and women with strengths and weaknesses, insights and biases. Though the scientist may work alone in the laboratory, much of science is also conducted in a social context, and this social context is just beginning to be seriously studied. The patterns that are emerging appear, in their own way, to be even more complex and difficult to interpret than those we find in diffraction photographs. The relation between a scientist's work and career has rarely been as fully documented as it is in the Wrinch Collection. It and the Symposium raise important questions which are sociological and philosophical as well as scientific and personel.

Dorothy Wrinch was born in Argentina in 1894 of English parents. She was educated at Girton College, Cambridge, studying pure and applied mathematics and the philosophy of science; by 1929 she had been awarded six advanced degrees, including the first D.Sc. ever awarded a woman by Oxford University. In addition to teaching, lecturing, and writing prolifically in these areas she was interested in the problems of professional women who are mothers, and even wrote a sociological treatise about them, under a pseudonym. During that period she was the wife of John Nicholson, an Oxford physicist. Her daughter Pamela was born in 1928. This marriage was dissolved in the thirties; in the forties she married O.C. Glaser, a biologist at Amherst College. Professor Glaser died in 1952.

Throughout the twenties her avocation was the problem of biological design. In the early thirties she decided to turn all her attention to it. She spent a year in continental Europe studying geometry in Vienna and chemistry in various laboratories; on her return to England she joined the discussion group called the Theoretical Biology Club, many of whose members later profoundly influenced the direction of modern biology. By 1935 Dorothy had suggested the hypothesis that the genetic code lay in the arrangement of amino acids; soon afterwards, she proposed the first model for globular proteins. Her model was based on the application of combinatorial mathematics to structural biochemistry. As it appeared to provide a simple explanation for a variety of experimental facts, it, and she, were quickly plunged into the international scientific limelight; the "cyclol controversy" which was to affect the rest of her life began almost immediately.

Dorothy Wrinch emigrated with Pamela to the United States just before World War II and lived here until her death. She continued to make contributions to the theoretical analysis of the diffraction patterns of complex crystals, and to many related subjects as well. (A complete bibliography of her 192 papers and books is included in Part II.) Although evidence continued to mount against it, in her own eyes her life work was the perfection of the cyclol model. In the sixties, she completed two monographs which she regarded as her scientific testament.

The cyclol controversy and its aftermath are and will be of special interest to historians and sociologists of science. The scientific questions which were debated centered on the existence of an arrangement of chemical bonds which her model required. They raise many fascinating epistemological questions about the nature of scientific knowledge and the relation between paradigm and experimental fact. The papers are also valuable for the light they shed on the personal questions that confronted her in the course of the controversy. In the history of science there have been many who fought for their theories and were proved right in the end, but there are also many, who fought equally well, whose theories were eventually proved to be wrong. "One knows very well that people sometimes tend to 'fall in love' with their ideas and find it very hard to give them up," writes Carolyn Cohen.[3] "When they are correct it is a sign of admirable perseverance and fortitude; when they are wrong, it is a pity. The problem is the judgment on the rightness or wrongness." In the Wrinch letters, we find every sort of advice and counsel, every shade of opinion, a wide range of motives. Reading them, we begin to understand how difficult it can be to make such judgments when one is in the eye of the storm.

The sociological aspects of the controversy are as well documented as the scientific and philosophical issues. The peculiarly acid vehemence of some scientific battles contrasts sharply with the image of objective observers searching for truth with the complementary tools of experiment and reason; it is surprising that such battles have not been the object of more study. If such a study is ever undertaken, the investigator will find the Wrinch papers an unusually rich source. Reading through the thousands of letters, manuscripts, proposals, reports, and reviews, we find a curious determination on the part of all concerned to reach final judgments before the evidence was in. In the thirties, only the roughest features of protein structures could be determined from x-ray diffraction data. The battle was "nail and tooth occasionally and not too nice," recalls Caroline MacGillavry.[4] "Everything about the structure of proteins was very uncertain at that time. Nobody really knew anything. This is why they fought so violently about it. It taught me this lesson that when people really know about things, they don't fight so viciously."

The controversy had profound and lasting depressive effects on Dorothy Wrinch's career, as other scientific controversies, past and present, have blighted the careers of many of those caught up in them. What drives the leaders of such crusades, and what inspires otherwise rational scholars to abandon objectivity and join them? A letter that Dorothy Wrinch wrote to a friend in 1940 is an interesting commentary on group behavior:

> "Did I tell you that a fellow who doesn't know me and doesn't know we are friends told a friend of mine about the conference at Gibson Island in August as follows (I was down for a paper on insulin but it is vilely hot there in Chesapeake Bay and I felt down and when there came a good excuse to go to Canada I just cut it.)
>
> Well he said 'Isn't it odd how people behave? I was with this group of men several nights in succession at Gibson Island and their one topic of conversation was working out how they would down Wrinch when she gave her paper. They were actually planning out all of the details of the attack on her. And then . . . they seemed like a lot of pricked balloons when she just didn't show up. . .'."[5]

Unhappily, much evidence suggests that scientific behavior has not improved in the intervening thirty-eight years.

It is natural to ask whether Dorothy Wrinch's career was fundamentally affected by the fact that she was a woman. In this era of increased awareness of the problems of women in science, her papers are a reminder that the question is complex. Many documents in the collection show, not surprisingly, that she would have had much greater success in obtaining funds and positions had she been a man. On the other hand we also find evidence of intellectual arrogance, strong ambition, a tendency to ignore conflicting evidence and a weakness for premature publicity — qualities which are welcomed in neither women nor men. On another level we find, explicitly and implicitly, philosophical arguments on the significance of form, the role of mathematical models in biology, and the nature of the barriers between scientific disciplines. These differences appear to have affected her career in subtle but significant ways. The changes in the structure of science which occurred during and after World War II, her emigration to America, and other external considerations undoubtedly also played their parts in her life. The relative weight to assign these many factors is difficult to judge. Dorothy herself was acutely aware of the difficulties of women in American science, but her letters show that in later years she believed that her views were primarily responsible for her scientific isolation.

The publication of these proceedings is particularly gratifying to those of us who were Dorothy's colleagues at Smith. During her years with us, she was an inspired teacher, a severe critic, an example of dedication and courage. We will always be grateful to her for the guidance and encouragement she showed to students and junior colleagues, and for the uncompromisingly high standards she set for herself and for others. We share her concern that the great questions of science be studied in their whole as well as in their parts. We extend this same concern to questions about science as a human enterprise. We hope that the present volume will contribute to this end.

All of the symposium speakers were friends of Dorothy Wrinch, in various times and places. The talks they chose to give in honor of her memory dealt with themes central to her work and life. The texts of their lectures, revised for publication, are the substance of Part I. In addition to the lectures there were discussions, in which not only the main speakers but also members of the audience participated, on the subjects of specialization vs. breadth, the relation between truth and beauty, views on scientific method, and the Wrinch papers. Transcripts of the first three discussions have been deposited in the Wrinch Collection; an edited version of the fourth appears in Part II, together with selected documents and papers from the Collection.

—*Marjorie Senechal*

# THE CONTRIBUTORS

*DAVID HARKER* was born in San Francisco, graduated from the University of California, and received his Ph.D. from California Institute of Technology. He taught chemistry at Johns Hopkins for several years and subsequently worked at the research laboratory of the General Electric Company. He was director of the protein structure project at the Polytechnic Institute of Brookyln for many years, and later served as head of the biophysics department of the Roswell Park Memorial Institute in Buffalo. He has published extensively on methods of crystal structure determination and on the structure of proteins. His current interests include the theory and applications of color symmetry. He is a member of the National Academy of Sciences.

*CAROLINE MacGILLAVRY,* born and educated in Amsterdam, received her Ph.D. with honors from the University of Amsterdam, where she was professor of chemical crystallography until her retirement. Her research has focused on crystal structure and its relation to chemical and physical properties of substances. She is an honorary Knight of the Netherlands Lion and a member of many professional societies, including the executive committee of the International Union of Crystallography. She was president of the board of the Child Guidance Clinic of Amsterdam for many years. Her latest book, *Fantasy and Symmetry,* is an analysis of Escher's periodic drawings.

*ARTHUR LOEB* was born in Amsterdam and graduated from the University of Pennsylvania. He received his Ph.D. in chemical physics from Harvard, where he is presently senior lecturer and head tutor in visual and environmental studies. He has taught physics at M.I.T. and for many years was staff scientist at Ledgemont Laboratory. He is an artist, a musician, and a dancer, as well as a scientist. His sculpture "Polyhedral Fancy" is in the permanent collection of Smith College; it is on display in Burton lobby. He is the author of *Color and Symmetry* and *Space Structures: Their Harmony and Counterpoint*.

*CAROLYN COHEN* was born in Long Island City, New York, and graduated from Bryn Mawr. She received her Ph.D. from M.I.T. She was a post-doctoral fellow at Kings College, London, and subsequently held positions at the Children's Cancer Research Foundation in Boston, M.I.T., and Harvard Medical School. She is presently professor of biology at Brandeis University, where she is chairman of the graduate biophysics program, and a member of the Rosenstiel Basic Medical Science Research Center. She has published many articles on the fine structure of muscles, and the structure of muscle proteins.

*RUTH HUBBARD* was born in Vienna, Austria, and graduated from Radcliffe College. After receiving her Ph.D. from Radcliffe and studying abroad, she joined the Harvard faculty, where she is presently professor of biology. She has made many contributions to our understanding of biochemistry and photochemistry of vision in both vertebrates and invertebrates. She was the 1975 Five-College Chemistry Lecturer. In recent years she has worked, written and lectured on a variety of health issues and on the politics of health care. At present a major focus for her is the sociology of science, particularly from a feminist perspective.

*GEORGE WALD* was born in New York and was educated there, graduating from New York University. He received his Ph.D. from Columbia University, and after a two-year post-doctoral fellowship (during which he first identified vitamin A in the retina) joined the faculty of Harvard University where he was Higgins Professor of Biology. He has received many awards for his work on the role of vitamin A in vision, including the Nobel Prize in physiology and medicine in 1967. He is a member of the National Academy of Sciences and the American Philosophical Society, and holds many honorary degrees.

# I. The Lectures of the Symposium

# COLORED LATTICES

David Harker

*Part of my privilege this afternoon is to reminisce a little bit about my acquaintance with Dorothy Wrinch — 1939-1940 it was. That is a long time ago, and I will tell you how it came about.*

We were all talking then about the notion which had recently been put forward that proteins, globular proteins, the ones that dissolve, occurred in molecular weight classes, approximately. That is not true — they don't. But it just happened in those days that there were thought to be classes of about 12 thousand, and about 40 thousand, of units we now call daltons. And at the same time as this, we began hearing that there was a Dr. Wrinch — Dr. Dorothy Wrinch — of Oxford, who had taken the then known chemistry of proteins and was able to construct models, based on the stereo-chemistry that was developing in those days, composed of hexagons in fabrics which would cover nicely tetrahedral and octahedral surfaces and would form closed figures for molecular weight classes which were not far from the then thought-to-be classes of molecular weight. So everybody was taking these molecular weight classes and the Wrinch hypothesis for the structure of proteins very seriously. Among the people who were doing that was Dr. Irving Langmuir of General Electric, and Irving Langmuir also was on the advisory board to the Chemistry Department at Johns Hopkins, where I was teaching chemistry and working in electron and x-ray diffraction, with a gas tube and a home-wound transformer — the kind of thing we did in those days.

Langmuir visited my laboratory and talked to me about a structure that I had been working on — acetamide, a structure very like a most elementary protein. It had a methyl group, a carbonyl group, an amino group, and nothing else, and so was related to the peptides, long polymers of amides of which proteins are composed, and it indeed turned out that the oxygen in the neighboring molecule was near the nitrogen along a direction where a hydrogen would have been found, if we could find hydrogens in those days. We also talked about the structure of protein, and Langmuir decided that I was the person to talk to a Dr. Dorothy Wrinch who was, I believe, here in 1939, early spring, consulting, giving lectures. He invited me up from Baltimore to Jones Beach, and himself from Schenectady along with Dorothy Wrinch, and we had a lunch in bathing suits on Jones Beach. And there was Dorothy Wrinch sitting with a leather attache case full of diagrams and little models, and Langmuir was talking madly about surface films of barium stearate and calcium stearate. I had an awful time diverting him. Finally, I said, "Dr. Langmuir, I am fascinated by these surface films, and I would like to know more, but here you have brought Dorothy Wrinch to talk to me about the structure of proteins, and what do you think we should do?" So he said, "All right. Let's talk about proteins." He was very definite. Remember him, any of you?

We then talked about proteins. Now Dr. Dorothy Crowfoot, who is at present Dr. Dorothy Hodgkin, had taken some x-ray diffraction pictures of insulin, a very low molecular weight protein, and it crystallized rhombohedrally, that is to say the symmetry of the crystal had a 3-fold axis in it. You could see three similar groupings in the x-ray diffraction pattern apparently related by rotations of 120° about such an axis. There was a method of treating the x-ray diffraction data which would show densities of interatomic vectors, and it was a picture of this that Hodgkin had published; it showed the 3-fold axis. The model of Dorothy Wrinch's protein also had a 3-fold axis, and the dimensions were about right (as they had to be because of the density and the molecular weight) for the model almost to fit the interatomic vector pattern — the Patterson function — that Hodgkin had published, and we argued back and forth about this. So I said, "Well, it is quite possible that this pattern and the Dorothy Wrinch structure for insulin are in agreement, but the resolving power is very poor, and there may be a large number of other structures that will fit this also. And anyway, this is a rhombohedral crystal, and anything that crystallizes rhombohedrally will have a 3-fold axis." So, after a while, we dispersed, I back to Baltimore, Langmuir to General Electric Research in Schenectady, and Dorothy Wrinch — I don't know just where she went at that moment.

I went back to Baltimore and told my department head, Donald Andrews, about the talk, and that it might be interesting if we could get support to work on the structure of protein. First thing I knew, Andrews had invited Dorothy Wrinch to spend the year at Hopkins, and we worked closely together. She taught a very nice course on the mathematics connected with organic structural chemistry, and she, of course, told us all about her "cyclols", as she called her models for protein molecules. We had an interesting time together, and I kept saying to her, "But why don't you learn x-ray diffraction techniques? You certainly understand the theory, and because other people have different interests and different responsibilities, if you really want to prove that your structure is right, you must learn how to prove it experimentally." She said, "Oh, I can't do that. I never can handle anything in the laboratory. Somebody else will have to do it. Won't you please do it?" And so it went.

We had lunch together in the attic of Remson Hall, the Chemistry Department at Johns Hopkins, and we would talk about all kinds of things, the goings on in Europe, Hitler invading Poland, Hitler invading France. "Oh," she said, "isn't it *awful!* We all thought that Churchill was such a horrible man, and he was right all the time!" And after a while she wangled an invitation from Eleanor Roosevelt to lunch at the White House in favor of rights for women of some kind, and she described her lunch at the White House, and how they all were seated, and Mrs. Roosevelt came in and said, "I'm so sorry, ladies, but Franklin is too busy to come to lunch today. Will you please accept his apologies?" So she didn't see Franklin Roosevelt after all. Sometimes she played the piano. She could read music very rapidly and very athletically. She was very gay and energetic. Charlie Chaplin's film, "The Great Dictator," came to Baltimore. She said, "Dave, call up home. Let's grab a sandwich and go see 'The Great Dictator' — the first show, at 7 o'clock." I said, "I can't do that. Katherine has got to feed the babies, and put them to bed, and get a babysitter. Why don't you come to supper at our place and we'll go to the 9 o'clock show?" And she resented that very much, because when she wanted to do something, she wanted to do it right away!

She was engrossed always with this hypothesis of the cyclol and was continually trying to gather more and more evidence for its plausibility and actual existence. I wrote a paper with her on bond energies which showed, as of that date, that it was impossible to determine on the basis of energetics whether the cyclol or the linear structure for polypeptides was the correct one. Pauling wrote some very derogatory articles which I think he should not have, and Dorothy Wrinch, who, as you know, was a great feminist, said, "Ah, it's the Y chromosome. If I had the Y chromosome, people wouldn't talk to me like this in public journals."

Well, eventually, I had to go to General Electric, and she went here to Smith, and we had less and less contact. Then came the great breakthrough: the structure of myolglobin, and the structure of hemoglobin, in 1960. I met her at Cambridge in England where Kendrew and Perutz were showing their models. And there was Dorothy Wrinch, chatting gaily with people, and the models were on a table about 10 feet behind her. I said, "Oh, hello, Dorothy. So the structure is now really known, and there is the model. Do you want to look at it?" She said, "No, I do not want to look at that model!" And she really didn't look at the model, until much later. She was still hoping that it was wrong.

Of course, now there are some dozens of protein structures known, and not one single one of them turned out according to her cyclol hypothesis. I am afraid we have to say that this hypothesis has been completely disproved. But it was a noble attempt, and it might have been right. All the arguments in favor of it and against it would have run exactly the same way in those days. My belief is, knowing what I do about Dorothy Wrinch, that she thought that if she had had a Y chromosome, then her structure would have been right.

Her value to me, to all of us, is that her hypothesis produced so much interest in the structure of protein that now, about 40 years later, there are close to 100 structures known. She certainly started me on protein structure. I was not really trying to prove or disprove her, I was trying to find out. (I think some of the people in England were trying to disprove her model.) And after 4 or 5 protein structures had emerged and turned out to be based on extended polypeptide chains arranged as in the alpha-helix of Pauling or the $\beta$-structure, similar to silk, (these seem to be the main structures present now) the question concerning Dorothy Wrinch's structure is academic, but it was a great attempt, produced tremendous scientific activity, and I think we owe a great deal of gratitude to anybody who can start the acquiring of such a vast body of knowledge.

Now I will get down to my business of the colored lattices. At the end of my involvement with proteins, it came about that there were huge protein structures, like the capsules of viruses, which are made up of subunits, all of which appear to be alike. They are arranged in similar ways for different viruses. This arrangement very often is based on the icosahedral point group 532, and virus after virus, which cause different diseases, seem to have similar structures. We did know that molecules of the same size, shape, and force field, approximately, could replace one another in structures whether they were chemically identical or not. It appeared to me that perhaps these viruses only appeared to be the same, to have the same molecular structure, but were made up of some subunits which might differ slightly chemically but not enough to spoil their geometrical and force-field properties. You would have a great many different possible virus particles with about the same appearance under the electron microscope, the same symmetry, and the same x-ray diffraction pattern. That started me to think about colored symmetry, and part of the motivation for what I am saying today came from that state of mind.

Today it's colored lattices I shall talk about. What is a lattice? A lattice is a set of objects which are all the same and distributed periodically. Figure 1 shows an example in 2 dimensions. If you imagine other similar layers out toward you, this collection would be a small sample of a three-dimensional lattice. This particular lattice consists of circles. I have distributed periodically white ones, slashed ones, and black ones. Their "color" is the only difference between

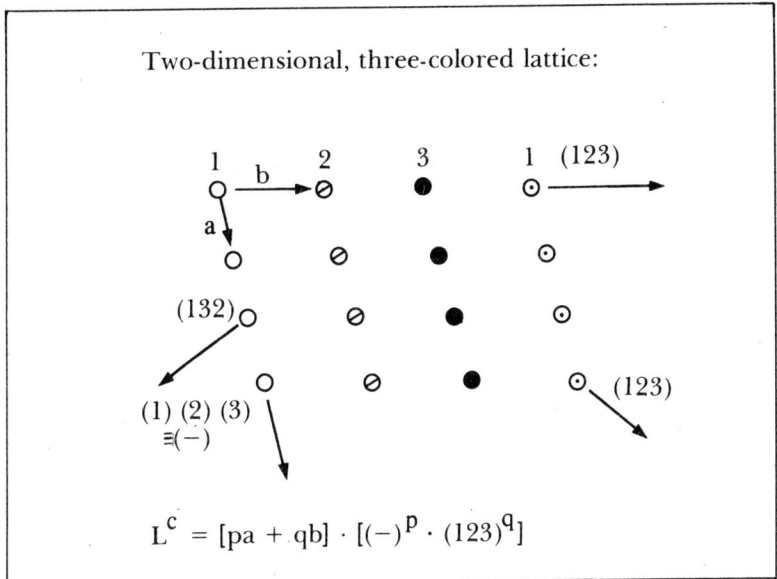

**Figure 1.** Colored symmetry operations $L^c$ of a three-colored net — a two-dimensional lattice.

them. They have the same size, the same shape, and, if you are color-blind, the same surroundings. The symmetry property is this: if I move along a vector of length **a** I arrive at a point exactly like the one I left. If I move the lattice by a vector of length **b**, the white ones go where the slashed ones used to be, the slashed ones go where the black ones used to be, and the black ones go where the white ones used to be. (I call these type 1, type 2, type 3, in the figure.) A motion of the lattice by a vector **b**, must be accompanied by a permutation (123) in order to restore the lattice to its initial state. A mathematician will recognize (123) as meaning "1 is replaced by 2, 2 is replaced by 3, and 3 is replaced 1." If I move the lattice by a vector **a**, 1 goes to 1, 2 goes to 2, 3 goes to 3, that is to say, nothing happens to the colors. I've symbolized that by $(-)$. If I move the lattice by multiples of $-$**b**, then the successive permutations are (132), (123), and so on. If I have p and q as integers and **a** and **b** are vectors in the plane of the figure, then a symmetry operation, a colored translation, $L^c$, is as given in Figure 1. This means, if I move the lattice by p**a** + q**b**, then I must apply this combined permutation, $(-)$ p-fold, and (123), q-fold. In 3 dimensions we have a more general case, (fig. 2). The symmetry operation is shown in this figure. In it are integers p, q, r, and vectors **a, b, c,** and there are permutations belonging to the motions called $P_a$, $P_b$ and $P_c$. This expresses the symmetry operations of a colored lattice in three dimensions.

---

Translation Group is

$$L = pa + qb + rc$$

Colored translation Group is

$$L^c = [pa+qb+rc] \cdot [P_a^p \cdot P_b^q \cdot P_c^r]$$

$P_a$ = the color permutation

required to restore the original arrangement of colors after moving the lattice by a, etc.

---

**Figure 2.** The geometrical translation group **L** and the colored translation group $L^c$ of a colored lattice.

You know that vectors, when they add, have the commutative property. If I go a distance p**a** (figure 1) and then a distance q**b**, I arrive at a certain point, but I also get to it if I go the distance q**b** first and the distance p**a** last. And that is commutation: p**a** + q**b** is equal to q**b** + p**a**. The permutations that accompany p**a** and q**b** will have to commute also to make things consistent. Permutation A followed by permutation B must be equal to permutation B followed by permutation A, etc. (fig. 3). We find that this isn't always true of permutations. Take A to be one permutation, and B to be the other one, and let us perform the two operations, A·B and B·A, (fig. 4). Let's do it this way: A·B means "1 goes to 2, 2 goes to 1," so 1 goes to itself; this is not noted. Start with 2; 2 goes to 3, then 3 goes to 4, and after that 4 goes to 1, 1 goes to 2. And so A·B = (234). You can verify, if you want, that the reverse order B·A doesn't give the same result. You

---

$P_a$, $P_b$, and $P_c$ must all cummute, i.e.:

$$P_a \cdot P_b = P_b \cdot P_a$$

$$P_a \cdot P_c = P_c \cdot P_a$$

$$P_b \cdot P_c = P_c \cdot P_a$$

---

**Figure 3.** The color permutations associated with lattice translations in the colored symmetry operations of colored lattices must all commute.

---

AB = (1234) (12) = (1) (234) = (234)

BA = (12) (1234) = (134) (2) = (134)

Thus, AB ≠ BA in this case;

A and B do not commute!

---

**Figure 4.** In general, permutations do not commute. Those appropriate for inclusion in colored symmetry operations make up a small subset of all permutations.

can see it without going into all the details; thus: 1 goes to 2, 2 goes to 3, which gives you 1 goes to 3 in B·A, which doesn't occur in A·B. So in general permutations do not commute, and we must leave out combinations of this kind.

Let's go back to Fig. 1. If I am color-blind, the white circles and the slashed circles look the same, but if I can see the colors, there is a one-colored lattice which has a reach 3 times as long in the **b** direction and the same length in the **a** directions as the one I see when color-blind. I will call this the one-colored lattice; while the one with the smaller reach is the color-blind lattice. If I call the color-blind lattice **L** and the one-colored lattice **L'**, then the vectors **a'**, **b'**, **c'**, characterizing the one-colored lattice are related to those of the color-blind lattice by a set of equations like the ones shown in fig. 5. They give the one-colored lattice's basis vectors in terms of those of the color-blind one. The array ($t_{11}$ $t_{12}$ $t_{13}$/$t_{21}$ $t_{22}$ $t_{23}$/$t_{31}$ $t_{32}$ $t_{33}$) is called

---

One-colored lattice **L'** in terms of "color-blind" geometrical lattice **L**.

If $\mathbf{L} = m\mathbf{a} + n\mathbf{b} + p\mathbf{c}$,

then $\mathbf{L'} = m'\mathbf{a'} + n'\mathbf{b'} + p'\mathbf{c'}$,

with $\mathbf{a'} = t_{11}\mathbf{a} + t_{12}\mathbf{b} + t_{13}\mathbf{c}$

$\mathbf{b'} = t_{21}\mathbf{a} + t_{22}\mathbf{b} + t_{23}\mathbf{c}$

$\mathbf{c'} = t_{31}\mathbf{a} + t_{32}\mathbf{b} + t_{33}\mathbf{c}$.

Here m,n,p,m',n',p', and all $t_{ij}$ are integers.

All this is symbolized thus:

**L/L'**($t_{11}t_{12}t_{13}$/$t_{21}t_{22}t_{23}$/$t_{31}t_{32}t_{33}$)

---

**Figure 5.** The colored lattice must contain some nodes all carrying the same color. If the colored lattice is to possess colored translational symmetry, the nodes of the same color must lie on a subgroup lattice **L'** of the geometrical, "color-blind" lattice **L**.

the matrix of the transformation from the color-blind to the one-colored lattice. The situation where there is a color-blind lattice **L** and a one-colored lattice **L'** derived from it by the matrix is symbolized in the last lines of fig. 5 and fig. 6.

---

The determinant of the $(t_{ij})$ matrix:

$$\Delta = \begin{vmatrix} t_{11} & t_{12} & t_{13} \\ t_{21} & t_{22} & t_{23} \\ t_{31} & t_{32} & t_{33} \end{vmatrix}$$

$|\Delta|$ is an integer equal to the number of colors in

$\mathbf{L}/\mathbf{L}'(t_{11}t_{12}t_{13}/t_{21}t_{22}t_{23}/t_{31}t_{32}t_{33})$

---

**Figure 6.** The number of nodes of the "color-blind" lattice **L** divided by the number of nodes of the "one-colored" lattice **L'** is given by the magnitude $|\Delta|$ of the determinant $\Delta = |t_{jk}|$ of the matrix $(t_{jk})$. This $|\Delta|$ is the same as the number n of different colors in the colored lattice $\mathbf{L}^c$.

Here are the vectors characterizing some orthorhombic lattices that I am going to talk about. In figure 7, O means orthohombic, P means primitive, I means body-centered, and F means face-centered. The 222 means that there are at least three 2-fold axes in mutually perpendicular directions. Look at the axes marked 222 OI. Ordinary body centering would mean you go out to the center of a rectangular cell in the shape of a parallelepiped and put another lattice point in the middle just like the ones at the corners. But this cell would spoil my matrix algebra, because the point in the center has fractional coordinates. So instead, any orthorhombic body-centered lattice will be based on vectors arranged like the ones marked 222 OI. First choose a lattice point, then choose another one derived by a 2-fold rotation from the first, finally a third one derived by using a different 2-fold axis. Vectors **a**, **b**, **c** from the origin to these describe a lattice with a primitive cell, identical with an ortho-rhombic lattice with a point in the middle of each parallelepiped. You can show in the same way that the all-face-centered ortho-rhombic lattice can be based on a skew set of vectors like that marked 222 OF, but the 222 OF cell will be primitive. Figure 8 shows axes for

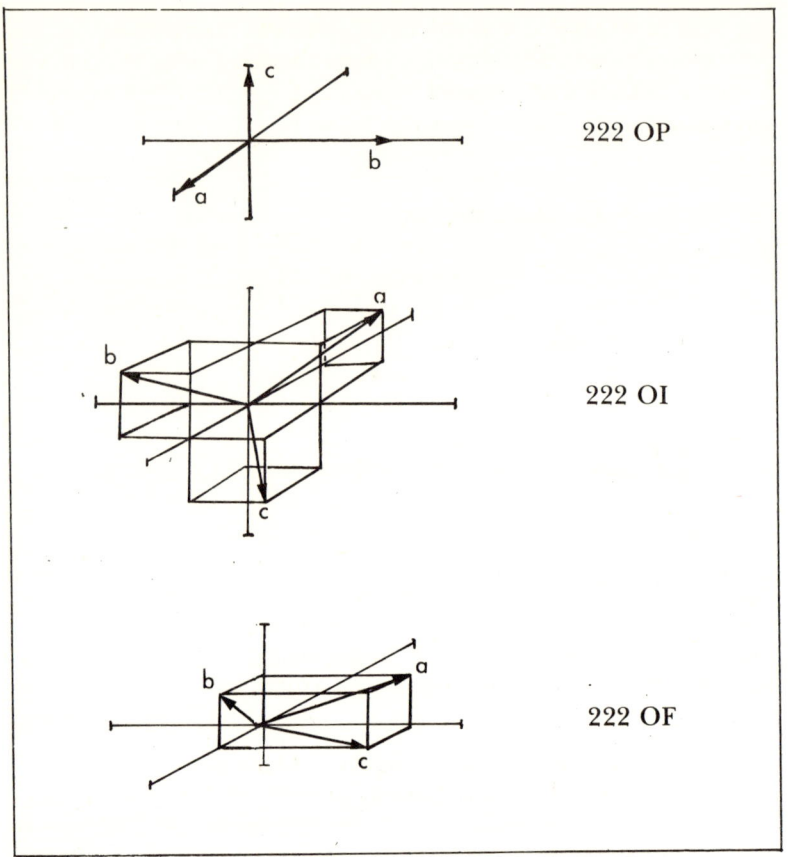

**Figure 7.** The most convenient triples of vectors for defining the primitive P, body-centered I, and all-face-centered F orthorhombic O lattices, so as to provide them with primitive unit cells, i.e., all lattice points have integral coordinates.

primitive cells producing C-centered, B-centered, or A-centered orthorhombic lattices. In OA, OB, and OC, the primitive cell is defined by two vectors of equal length, bisected by a 2-fold axis, and at right angles to another vector along a second 2-fold axis. Figure 9 shows tetragonal primitive cells. S means square; SP means primitive tetragonal, and SI means body-centered tetragonal.

Figure 10 is a table of the possible orthorhombic colored lattices. I am going to assume that the one-colored lattice and the color-blind lattice are based on the same crystal system. There are then seventeen types of colored lattices in the orthorhombic system. The first

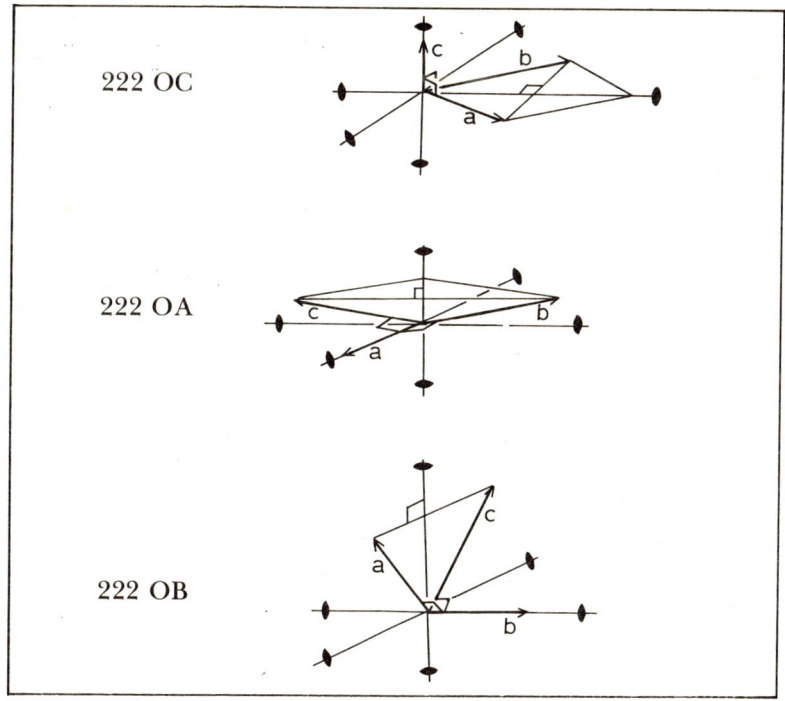

**Figure 8.** Vector triples for conveniently defining the C, A, and B end-face-centered orthorhombic O lattices.

column of the table gives the centerings of the color-blind and one-colored lattices and the matrix relating the latter to the former. For instance, OC/I(pqr/$\bar{p}\bar{q}r$/qp$\bar{r}$) means the color-blind lattice is C-centered orthorhombic, and the one-colored lattice is body-centered; the most general transformation matrix for this case is in parentheses. The determinant of the matrix in column one give a possible number of colors. The integers p, q, and r define this matrix. In column two is the formula for the value of this determinant in terms of p, q, and r. This is the number of lattice points of the color-blind lattice in one cell of the one-colored lattice, and is equal to the number of colors present. Some centering combinations don't allow this number to be arbitrary. For instance, if the lattice is color-blind primitive, and one-colored body-centered, then the number of colors has to be a multiple of 4, and if both lattices are C-centered, the determinant has to be equal to $r(p^2-q^2)$, which means you can have a three-colored, end centered, orthorhombic lattice:

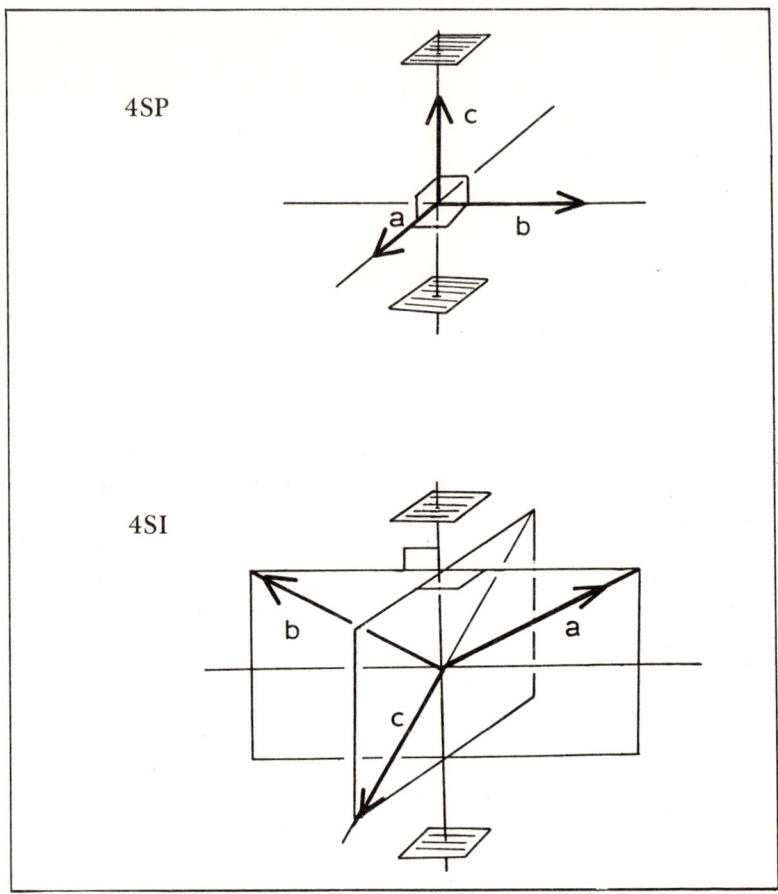

**Figure 9.** Vector triples for conveniently defining primitive P, and body-center I tetragonal S lattices.

when r is 1, p is 2, q is 1. I won't give you the whole list of colored lattices for all systems. If the list is complete, it includes the triclinic, monoclinic, orthorhombic, tetragonal, hexagonal and cubic cases, and there are 39 combinations.

Now I am going to show you a few of these lattices. One of the simplest cubic ones is the famous sodium chloride structure. Its axes are shown at the top of figure 11. The sodium chloride structure is built on a cubic array of atoms, and the color-blind lattice, if you can't tell the difference between sodium and chlorine, is based on 3 equal-lengthed orthogonal vectors **a**, **b**, and **c**. However if you can

Examples of Three Dimensional Colored Lattices for which the Geometrical (or Color-Blind) Lattice, **L**, and the one-colored Sublattice, **L'**, Belong to the Orthorhombic Crystal System.

| **Orthorhombic 222 Symbol** | **No. of Colors** $= |\Delta|$ |
|---|---|
| OP/P (p00/0q0/00r) | pqr |
| OP/C (pq0/$\bar{p}$q0/00r) | 2pqr |
| OP/F (0pq/r0q/rp0) | 2pqr |
| OP/I ($\bar{p}$qr/p$\bar{q}$r/pq$\bar{r}$) | 4pqr |
| OC/P (p$\bar{p}$0/qq0/00r) | 2pqr |
| OC/C (pq0/qp0/00r) | $r(p^2-q^2)$ |
| OC/A (p$\bar{p}$0/qqr/$\bar{q}\bar{q}$r) | 4pqr |
| OC/F (ppq/r$\bar{r}$q/p+r,p−r,0) | 4pqr |
| OC/I (pqr/$\bar{p}\bar{q}$r/qp$\bar{r}$) | $2r(q^2-p^2)$ |
| OF/P ($\bar{p}$pp/q$\bar{q}$q/rr$\bar{r}$) | 4pqr |
| OF/C (p$\bar{p}$q/q$\bar{q}$p/rr$\bar{r}$) | $2r(q^2-p^2)$ |
| OF/F (p$\bar{q}$q/$\bar{r}$,p−q+r,r/q−r,r−q,p+r) | $(q-p-2r)(q^2-p^2)$ |
| OF/I (pqr/q,p,−(p+q+r)/−(p+q+r),r,q) | $2(p+q)(q+r)(r+p)$ |
| OI/P (0pp/q0q/rr0) | 2pqr |
| OI/C (p,q,p+q/p,$\bar{q}$,p−q/rr0) | 4pqr |
| OI/F (p+q,q,p/q,q+r,r/p,r,p+r) | 4pqr |
| OI/I (pqr/q−r,p−r,$\bar{r}$/r−q,$\bar{q}$,p−q) | $(p+q-r)(p-q+r)(p-q-r)$ |

**Figure 10.** Table of the symbols of all seventeen possible orthorhombic colored lattices, and the allowed number of colors $n = |\Delta|$ for each. The symbols are derived from that at the bottom of Figure 5 by prefixing O, for orthorhombic. (See text for a fuller explanation of these symbols.)

distinguish sodium from chlorine, we must talk about the one-colored lattice, say the chlorine lattice; it is based on vectors **a'**, **b'**, and **c'**, which go to the face-centers of a doubled color-blind cube. We have Q for cubic, P color-blind, F one-colored. The matrix, as you see, is (011/101/110). If you evaluate the determinant, it is 2: two colors, sodium and chlorine. If you squeeze this structure along the cubic body-diagonal, it is not cubic any more, but it *is* rhombohedral: it still has the 3-fold axis, but it has lost most of its other symmetry. However, the same transformation matrix is available to the rhombohedral one-colored lattice, and the transformation will still have the determinant 2. There are examples of this structure: the

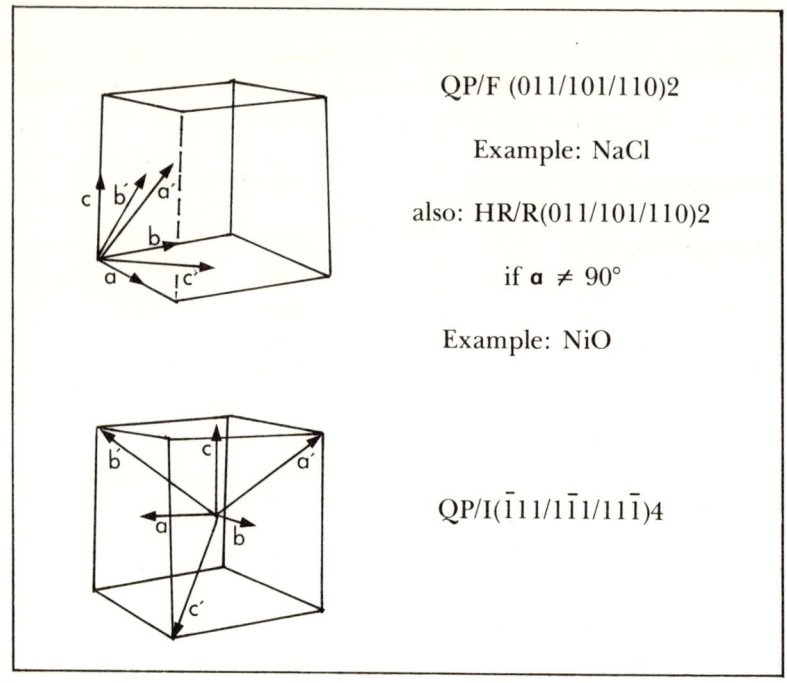

**Figures 11 & 12.** The axes of the "color-blind" lattice **a**, **b**, and **c**, and of the "one-colored" lattice **a′**, **b′**, and **c′** for four possible cubic Q colored lattices.

low-temperature form of nickel oxide. Another is that of the alloy CuPt shown at the bottom of figure 13. So this is a way of describing such simple structures.

(The four-colored cubic lattice with the axes shown in the bottom picture of figure 11 doesn't correspond to any real crystal structure I know about.)

Here is another one: AuCu. The top of figure 12 shows the axes, and the middle of figure 13 shows the corresponding two-colored lattice. (I put S for tetragonal, because T has other uses.) This lattice is square, body-centered color-blind, primitive one-colored. The determinant turns out to have a value of 2, and the ordered alloy AuCu has this structure. If c/a (for crystallographers) is $\sqrt{2}$, we have a cubic close-packing. This is the structure of a disordered alloy made up of atoms: half of them gold and half copper. If it is allowed to order, then the gold lattice is primitive tetragonal. The middle of figure 13 shows this structure. We have alternate layers of gold and

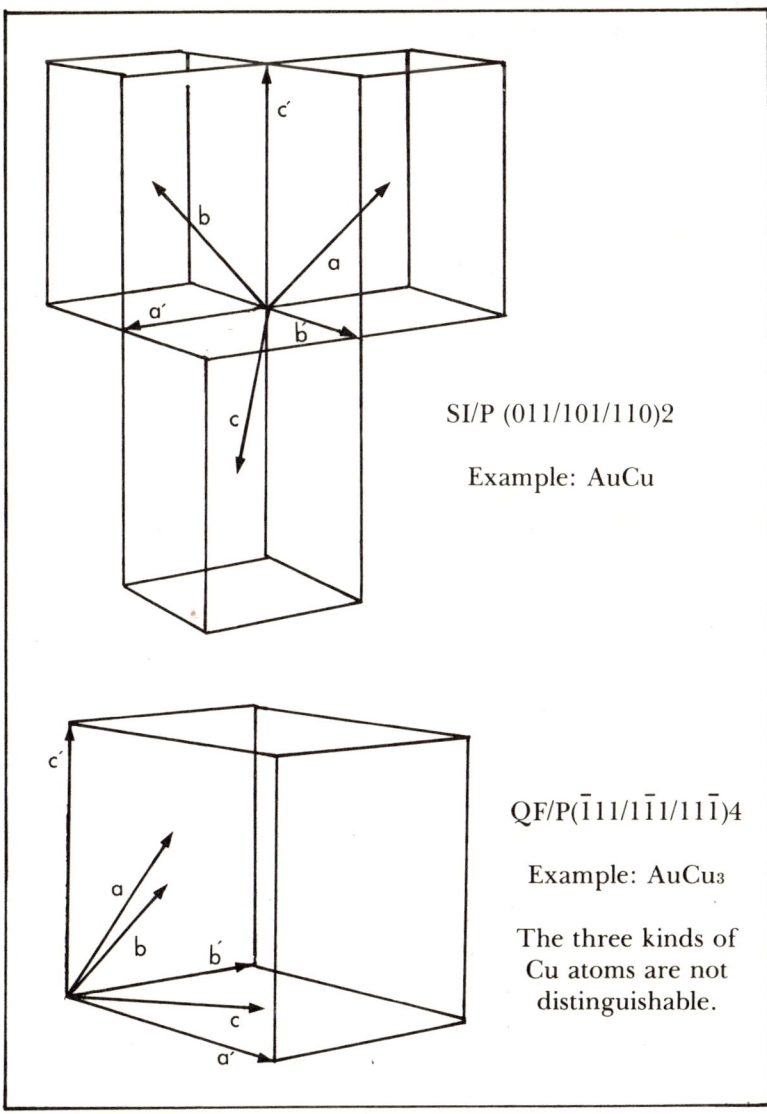

**Figure 12**

copper atoms in square-packing, perpendicular to a cube axis. In one layer you will have only gold, and in the next layer only copper, and so on.

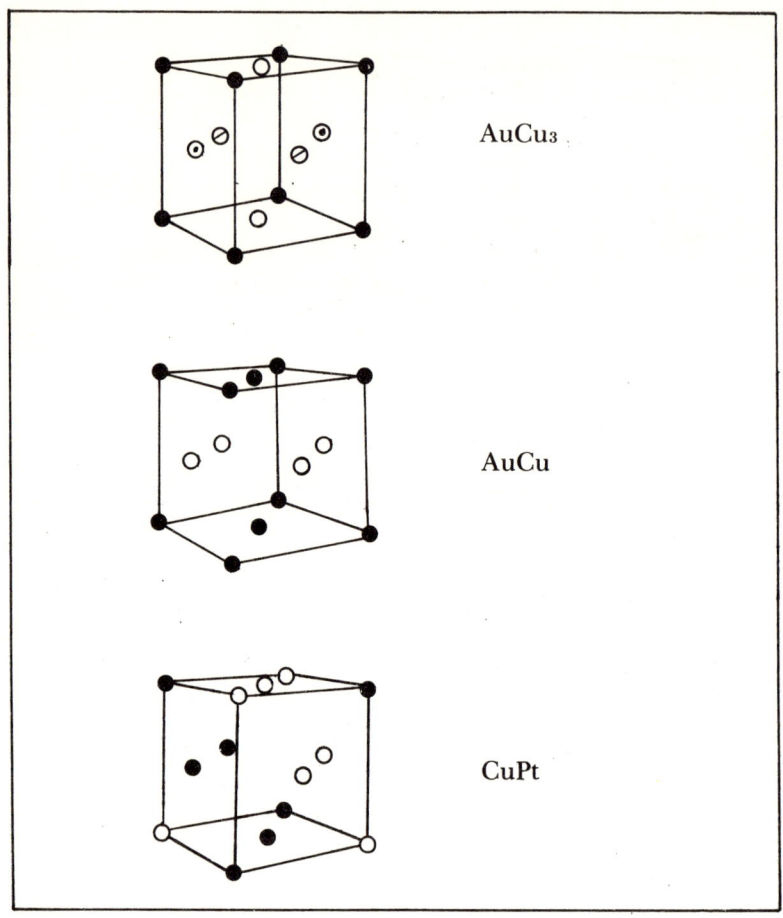

**Figure 13.** The atomic arrangements in the crystal structures of the alloys AuCu3, AuCu, and CuPt. The atoms are at lattice points. Open, closed, dotted, and slashed circles indicate different "colors" of lattice points. In AuCu3 the three copper atoms have different "colors", although they are the same chemically.

Here is another one: cubic, color-blind face-centered, one-colored primitive based on the axes in the bottom of figure 12, and the determinant is 4, so you have to have 4 different kinds of atoms present. AuCu3 has this structure. In AuCu3, three of the atoms have to be chemically alike (the coppers), but they are different crystallographically. In colored-symmetry crystallography, the three colors are different, and the gold is different from all of them. Here, at the top of figure 13, is a better picture of AuCu$_3$ with the three different coppers given different "colors": open, dotted, and slashed circles.

Here is a nice one: AlFe$_3$ (figure 14). It is body-centered cubic, if you are color-blind, and the one-colored lattice is face-centered. It accommodates 4 different kinds of atoms. Let's take the open, dotted and slashed atoms to be iron, and the filled atoms to be aluminum. We shall get the AlFe$_3$ structure in this way. This structure has been found, and a lot of other ones related to it: the iron can be substituted by various other things: — chromium, nickel, and what not. These alloys are often magnetic. And here is my own structure. I determined the structure of Ni$_4$Mo, when I was at General Electric (figure 15). It is tetragonal, and although this picture doesn't show it too well, the distance from front to back is $\sqrt{2}$ times the atom-atom distance vertically. It has a body-centered tetragonal color-blind lattice and the one-colored lattice is also body-centered tetragonal; the matrix is SI/I (201/101/012) and the

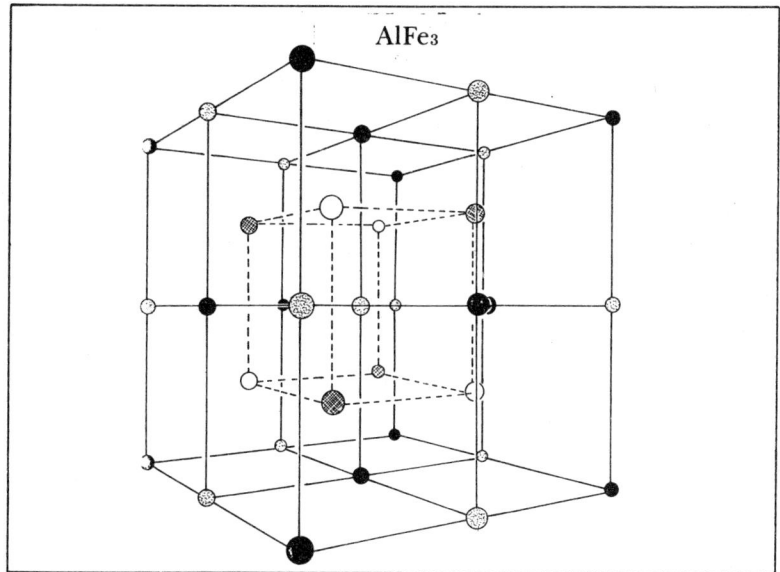

**Figure 14.** The atomic arrangement in the crystal structure of the alloy AlFe$_3$. The symbolism is the same as in Figure 13.

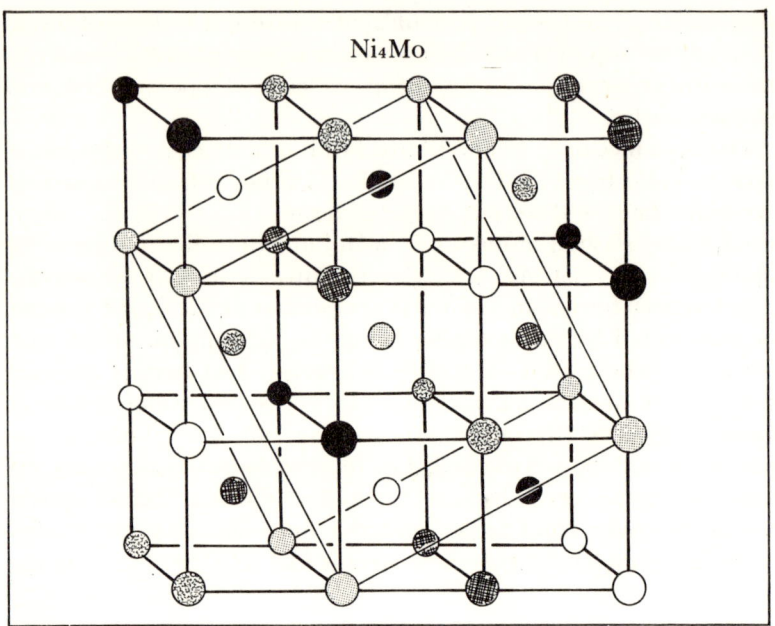

**Figure 15.** The atomic arrangement in the crystal structure of the alloy Ni₄Mo. The symbolism is similar to that of Figure 13.

determinant is 5. We thus have 5 colors of which 4 are given to nickel atoms, although they are chemically the same, while the 5th one is given to molybdenum atoms (which you might take to be the black).

I have shown you what I think are fairly pretty pictures of simple structures predicted on the basis of the colored lattice theory, but there are probably many other applications, too. Instead of having chemical nature as the color, you might have orientation of a magnetic vector belonging to an atom, in which case this theory applies to magnetic structures. Every electron has a plus or minus spin, and in fact this application has been made, using the two-colored lattices. But these aren't the only ones. There are atoms with a spin of 1, which are allowed, by quantum mechanics, to have plus, zero, and minus projections on a magnetic field, and so you could use 3 colors, and you could have 3-colored lattices, and so on. I won't labor this point, except that the theory applies to structures with spins.

Another thing we could do is to apply colored symmetry theory to large molecules, which have subunits that are not identical chemically, but have the same size, shape, and force-field (to the resolving power required), and you could build up ordered arrangements of subunits. Virus molecules may have such 3-dimensional structures, as I said earlier. We might to some extent describe the structure of collagen this way. It is a linear peptide of glycine, proline, and hydroxyproline, with some variations. This would be a very simple example of a three colored string of beads.

I don't know where all this will lead. This colored lattice theory is also the basis of colored space groups. In lattices, as I showed you, the repeating objects are all parallel, but you can also have 3-dimensional structures where things alternate orientation in various ways and in various directions. There are, for instance, what we call structures with screw axes where only every 3rd object is parallel to the 1st. If you apply colors to these and other such structures, you would have colored space groups. These would apply again to magnetic properties, to ordered crystalline compounds, etc. There is a lot of lovely mathematics involved. Mathematics is what I am going to do now until somebody tells me to stop.

*By the way, I think Dorothy Wrinch would have liked this stuff!*

# SYMMETRY IN THE THREE KINGDOMS, ANIMAL, VEGETABLE, MINERAL

Caroline H. MacGillavry

*The subject I am to talk about is one which was certainly dear to Dorothy Wrinch's heart. Of course it would have been better if she had been able to deliver this lecture herself, but it can't be done. So I will do my best to give you some ideas about the subject, and if I manage to do this, please think that this is done in honor of her memory.*

What is symmetry? Intuitively, we would say: it is a regularity in shape of an object or pattern so that, if only a suitable part of it is seen, the rest can be guessed, or "predicted". Take, for example, William Blake's famous poem:

> "Tiger, tiger, burning bright
> Through the forests of the night,
> What immortal hand or eye
> Did frame thy fearful symmetry?"

Let us analyze what Blake means by the word "symmetry". If we see only the left flank of a tiger, our experience from the zoo or from looking at our own tabby cat permits us to have a pretty good idea how the right flank of the tiger will look: we imagine it to be the *mirror image* of the left flank (figure 1). The position of this imaginary mirror is, of course, upright, and passing through the spine and the outstretched tail of the tiger. In this position the mirror interchanges neither back and front, nor above and below, but only left and right. That is, it inverts the *sense* of direction *perpendicular* to the mirror.

Such *mirror symmetry* is easily recognized by us in any figure or object, presumably because we possess it ourselves.

What happens if we look at ourselves in a mirror in front of us? We

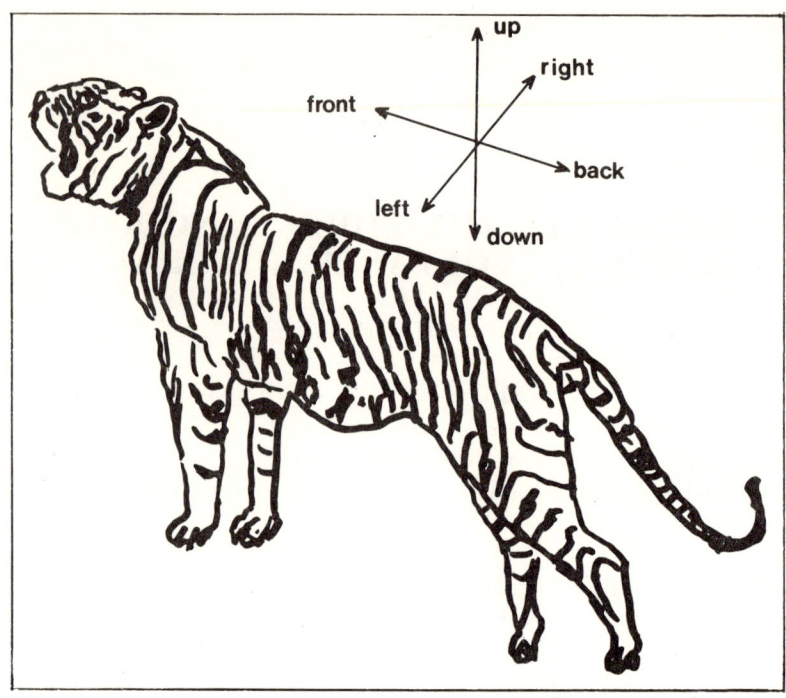

**Figure 1.** The left flank of a tiger.

*think* that again left and right are interchanged in our mirror image with respect to ourselves. But is this really so? An animal in the same position *sees* a rival confronting it, and will act accordingly (figure 2). The animal is right: all that a mirror in this position does is change the position of front and back, that is, again, the sense of direction perpendicular to itself.

Can we — not really, but in thought — coincide with our mirror image in the pier glass? Yes we can, *by making a half turn* about a vertical line in the mirror, where it is intersected by our own mirror plane. Such a *rotation* about a vertical axis interchanges back and front *and* left and right. Combined with the pier glass which interchanges back and front, the total result is an apparent change of left and right. This simple thought experiment contains one of the laws of symmetry: "Two mirror planes intersecting at right angles generate a rotation of 180° about the line of intersection."

Does our body in itself possess rotation symmetry? Of course it does not; only a full turn about any direction will make us look the

*Symmetry in Three Kingdoms* 33

**Figure 2.** An animal sees a rival in a mirror.

same as before. This may be the reason why playing-cards — kings, queens, valets which do possess rotation symmetry look so unnatural to us (figure 3). Rotating this card in its own plane there are, within a

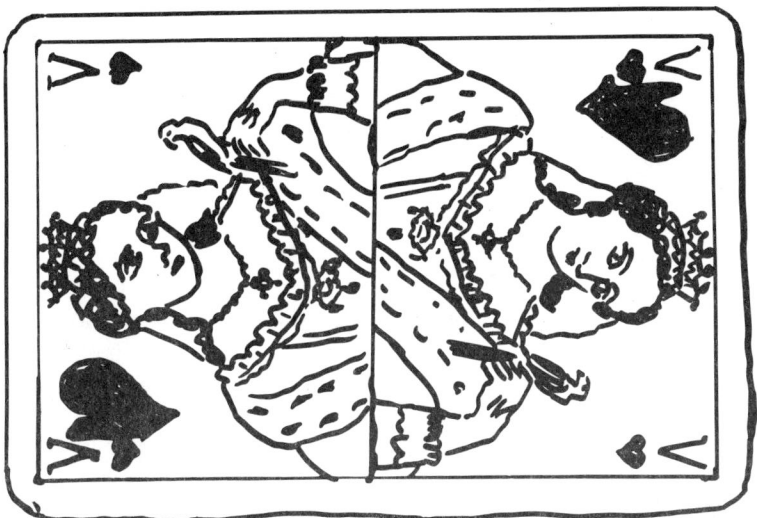

**Figure 3.** Playing card.

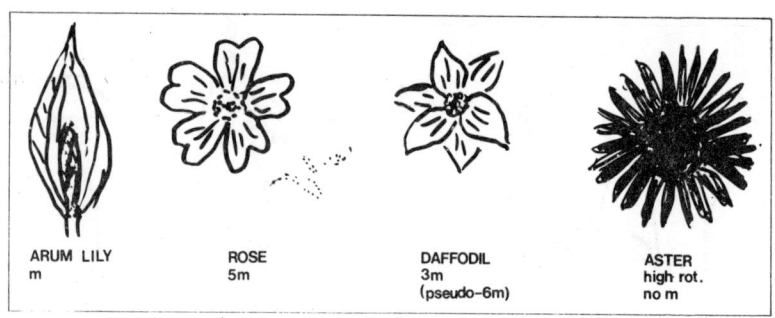

**Figure 4.** Flower symmetries.

full turn, two positions looking exactly the same. We call this *twofold rotational symmetry*.

Mirror symmetry as seen in humans and tigers is also apparent in many flowers such as snapdragons and irises. How about rotational symmetry? Again, in the vegetal kingdom it is very well known: the flowers of rose and forget-me-not have fivefold rotation symmetry; tulips and daffodils appear to be sixfold but are actually only threefold, their petals being alternately higher and lower. Aster, sunflower, daisy have high rotational, but no mirror symmetry (figure 4).

No so-called "higher" animal possesses rotational symmetry. An animal which reaches its aim — prey, mate, place to rest — by walking, hopping, flying or swimming cannot afford to have a rotation axis: it would not know which way to move! However, animals that are sedentary or at most crawl, or float about passively like plankton, do quite often have high rotation symmetry. Starfish are a well-known example of fivefold symmetry. Note that some are even tenfold! (figure 5)

More primitive forms of life quite often have high and complicated symmetry. In the 19th century, the German biologist Ernst Haeckel described and pictured the radiolaria. These are unicellular organisms which protrude their protoplasm in thin filaments through the pores of their cell wall, in order to collect what they need for subsistence. So both for safety and for stability they build themselves a house, an external skeleton made of silica (figure 6). These delicate, but remarkably strong constructions persist when the organism itself has died; they sink to the bottom of the ocean and are eventually found later in marine formations and in beach sands.

*Symmetry in Three Kingdoms* 35

**Figure 5.** Starfish. From F.A. Roedelberger, *Fauna Ferner Inseln*, Buchverlag Verbandsdruckerei, AG-Bern 1965.

Many of these skeletons have intricate and beautiful symmetry.

High rotational symmetry is also well known for some polyps, e.g. the sedentary sea-anemone, and from the floating jelly-fish (figures 7,8,9).

Sometimes animals which themselves possess nothing but bilateral symmetry aggregate to form a starlike figure of high rotational symmetry, in the same way as girls may do in figure swimming. This is known from various small organisms occurring in zoö-plankton (figure 10). Why do these creatures aggregate in this way? I do not know. Is it easier to float about, or are their most vulnerable parts better protected?

We have now seen examples of rotation about one axis and of mirror symmetry, apart or in combination. Are there bodies with rotation about more than one direction? I will not go into all details but return to the radiolaria. Figure 11 shows some skeletons with

36  *Structures of Matter and Patterns in Science*

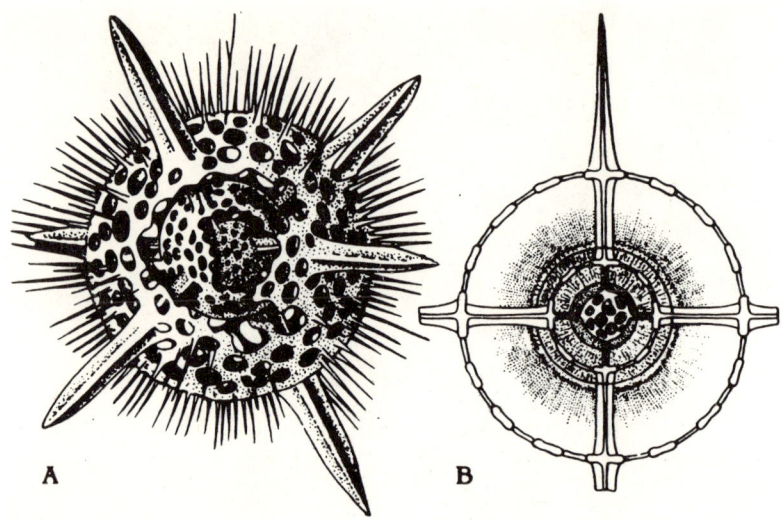

**Figure 6.** Radiolaria (after E. Haeckel).

**Figure 7.** Symmetry "m": Dorcadospyris donoceras. From Roedelberger, *op. cit.*

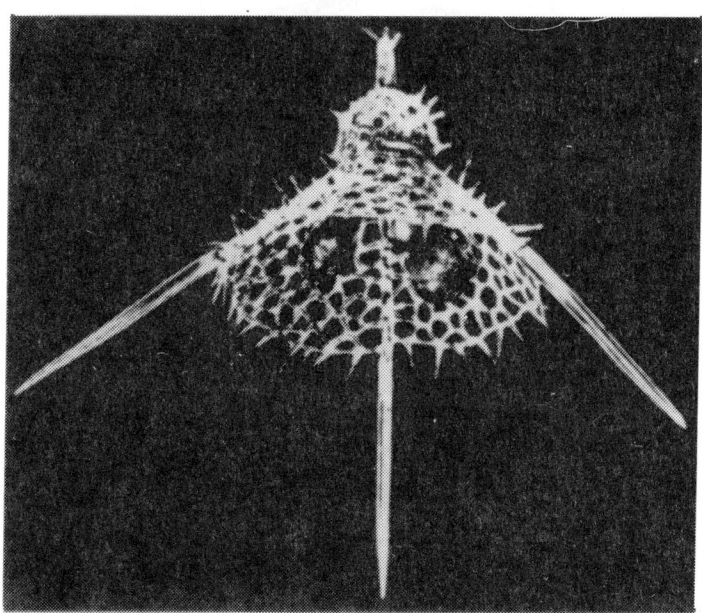

**Figure 8.** Symmetry "3m": Dictypnimus hertwigii. From Roedelberger, *op. cit.*

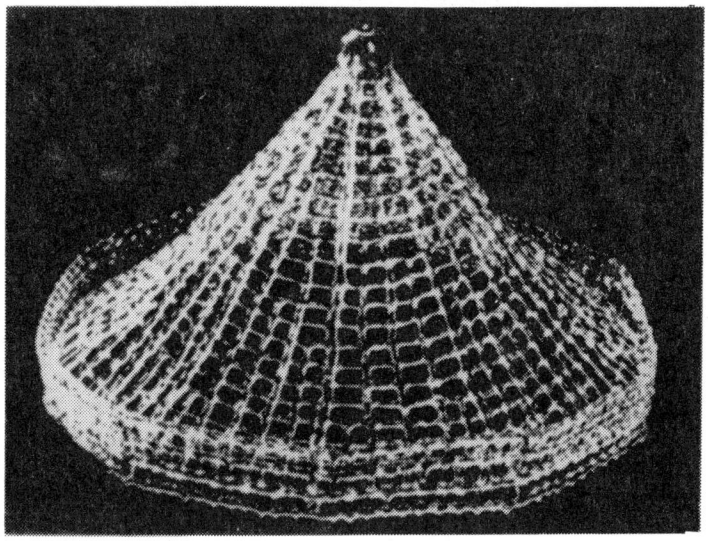

**Figure 9.** Symmetry "10m": Sethoformis eupylium, From Roedelberger, *op. cit.*

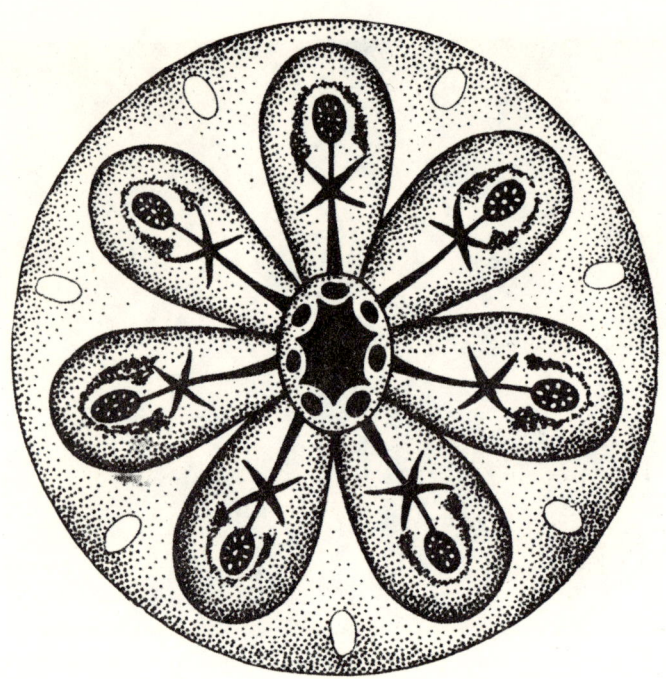

**Figure 10.** Zoö-plancton (after E. Haeckel).

high and complicated symmetry. We see spherical structures with more or less regularly arranged spikes and holes; we see an octahedron, two bodies with twenty regular triangular faces (icosahedra), and one having twelve regular pentagons (regular pentagondodecahedron). Except the sphere, all these bodies are so-called regular polyhedra, whose faces are equilateral polygons. Apart from those shown in figure 11, there exist two others: the regular tetrahedron formed by four equilateral triangles, and the cube with six square faces.

The total rotational symmetry of these regular bodies is described in the table below.

| Body | Faces | Rotation Axes |
| --- | --- | --- |
| tetrahedron | four regular triangles | four threefold three twofold |
| cube | six squares | four threefold three fourfold |

## Symmetry in Three Kingdoms

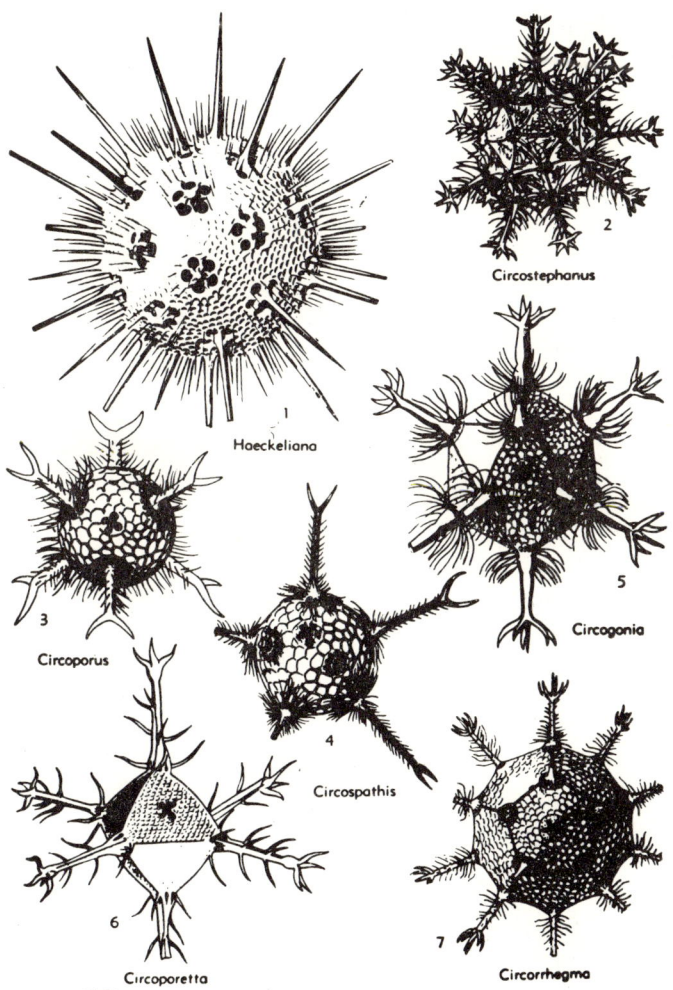

**Figure 11.** Isometric radiaria (after E. Haeckel).

| | | |
|---|---|---|
| octahedron | eight triangles | six twofold |
| pentagondodecahedron | twelve pentagons | six fivefold |
| | | ten threefold |
| icosahedron | twenty triangles | fifteen twofold |
| sphere | infinite number of infinitely small faces | infinite number of infinite-fold axes |

Does the sphere occur in our three kingdoms? Yes. Some examples are:

> pollen cells of some plants .............vegetable kingdom
> snail's eggs ............................animal kingdom
> oil drops in an emulsion ................mineral kingdom

Some bilateral animals, when in danger, roll up to simulate spheres; hedgehogs, or better still: isopods, a small kind of crustacea (figure 12). Figure 13 shows an example of a pentagondodecahedron-shaped coating of a small kind of alga. Organisms trying to protect a more or less isometric interior by an armor show a preference for either a sphere, or a polyhedron that approximates the sphere as well as possible in a body composed of flat surfaces. This appears to be why we don't often find tetrahedra or cubes surrounding living organisms.

On the other hand, in the mineral world, we find crystals whose natural growth form is the cube (figure 14), the octahedron or the tetrahedron, but *never* the higher symmetrical forms with fivefold symmetry. Nor do crystals ever have forms with rotational symmetry higher than sixfold. *Why is this so?*

It has long been suspected, i.e. from the way they grow from

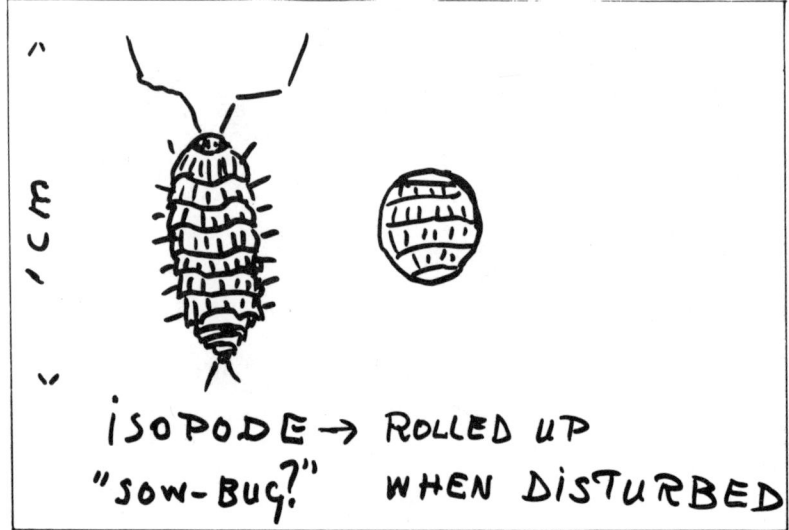

**Figure 12.** Isopod.

## Symmetry in Three Kingdoms   41

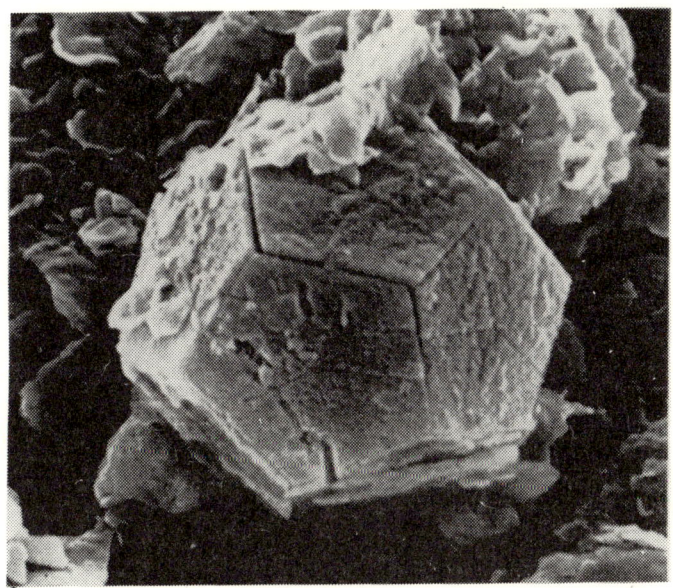

**Figure 13.** Coccolith armour.

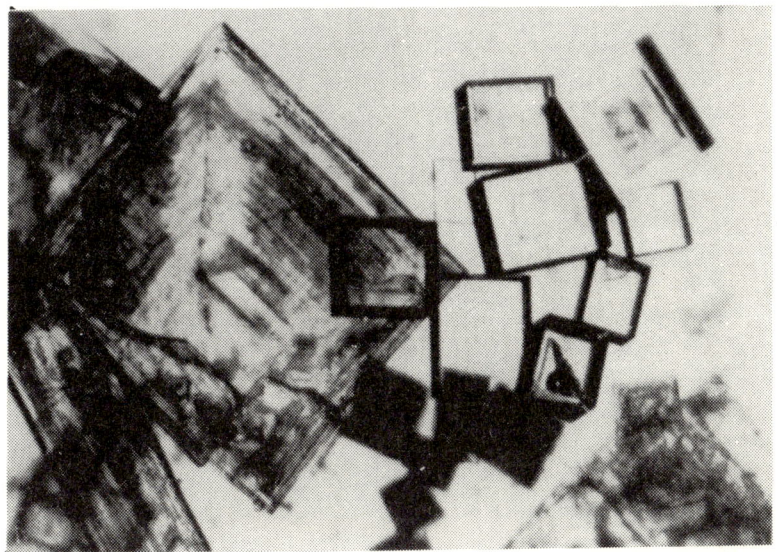

**Figure 14.** Cubes of rocksalt, NaCl. Photograph by A. Kreuger.

solution, and from the faces they develop naturally, that crystals are built up of *regularly repeating identical units*. Unlike living organisms, where life is sustained by a complicated interplay of a variety of chemical compounds, crystals are *homogeneous:* each species has a definite chemical composition throughout. These supposed identical units must be chemical units, atoms or groups of atoms, often chemically bound to form molecules. Atoms and molecules are very, very small, too tiny to be seen even under the best light microscope. Only special modern techniques, such as the so-called electron microscope, permit us to make a very much enlarged image of such molecules, and even then only of the larger among them, such as protein molecules made up of many thousands of atoms. Figure 15 shows the three-dimensionally regular stacking of these small particles, in a virus crystal, and how the faces of this tiny crystal are densely populated with identical units. We recognize also, in such a periodically repeated pattern, a new *symmetry element,* namely the

**Figure 15.** A virus crystal. Photograph by R.W.G. Wyckoff. Reprinted with permission from *Crystals: Their Role in Science and Nature* by Charles Bunn, New York, Academic Press, Inc., 1964.

Symmetry in Three Kingdoms 43

parallel shift over a constant distance in various well-specified directions. Such an operation is called a *translation*. From looking at the picture one can anticipate how the next layers of molecules will be added on the various faces when the crystal grows. Such anticipation was mentioned at the very beginning of my talk as an intuitive criterion of symmetry.

We are familiar with the concept of translation in the macroscopic world: brick walls, tiled floors, wallpaper, woven fabric and honeycombs are all examples of periodic patterns which repeat in a *plane*,

**Figure 16.** Trees reflected in the Amstel river. Photograph by James L. Amos.

that is, in two dimensions. In crystals, the units repeat in three-dimensional space.

Can translations combine with other symmetry, mirror and rotation? Let us start with patterns that repeat only in one direction. In my country, trees are often planted at regular distances along a country road or along a canal. In figure 16 you see the trees reflected in the water. So the total symmetry of the tree pattern is translation in one direction and a mirror plane parallel to that direction. Mirror symmetry can also be perpendicular to the single translation. It then repeats with the translation; moreover, a second set of mirror planes comes halfway between the first set (figure 17). Alternatively: two

**Figure 17.** A squirrel between two mirrors.

parallel mirrors generate a translation normal to the mirrors. This is experienced by a lady trying on a new dress in a dressing closet with mirrors on two opposite walls. Figure 18 shows two examples of band ornaments where the *one* mirror parallel to the translation is combined with the two sets of repeated mirrors perpendicular to it.

Let's go back to the animal kingdom. What happens to your own symmetry plane when you start walking or climbing? A picture of one of our nearest relatives among the vertebrates (figure 19) shows that its original symmetry is lost, but its hands and feet are still each other's mirror image, although now one is higher than the other. The track of the tiger (figure 20) still betrays its "fearful symmetry", but also the way it has moved while walking. In the track you find the imprint of the right paws one step further than their mirror image, the track of the left paws. After *two* steps, the pattern is repeated: a translation is generated. So the original mirror plane of the tiger

**Figure 18.** Band ornaments in a French Cathedral. From Dr. W. Bronkhorst, *Van Moissac tot Reims*, W. Bergmans, Tilburg, 1946.

**Figure 19.** Tarsius syrichta.

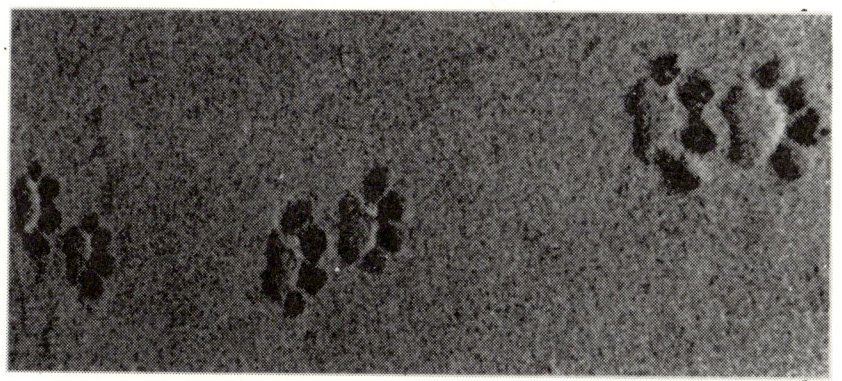

**Figure 20.** A tiger track.

becomes a *glide plane* in the track! This glide mirror symmetry is often found in linear ornaments in architecture or embroidered bands (figure 21).

**Figure 21.** Band ornaments with glide symmetry. From A.V. Shubnikov and V.A. Koptsik, *Symmetria v Nauke i Iskusstve*, Moscow 1972.

When we look more closely at the first two examples in figure 21, we see there is something unnatural about them. They are clearly stylized from plants. But it looks as if the plants were first put on a piece of blotting paper and their leaves and flowers flattened out so that they all look towards you, as you do when you put a plant in a flower press to dry. Naturally, leaves develop on a plant stalk as along a screw, or rather, as along a winding staircase. Each new leaf sits one step higher than the last, and it is also *rotated* about the stalk over an angle which for a given plant is said to be fairly constant. So in reality we have in such plants a combination of translation with rotation. The stalk with its leaves is said to have a *screw axis*.

Many papers have been written about this regular arrangement of leaves, and many theories put forward. The authors complain that it is difficult to measure and check the angles between subsequent leaves. So I measured my own dragon tree *(Dracaena fragrans)*. On

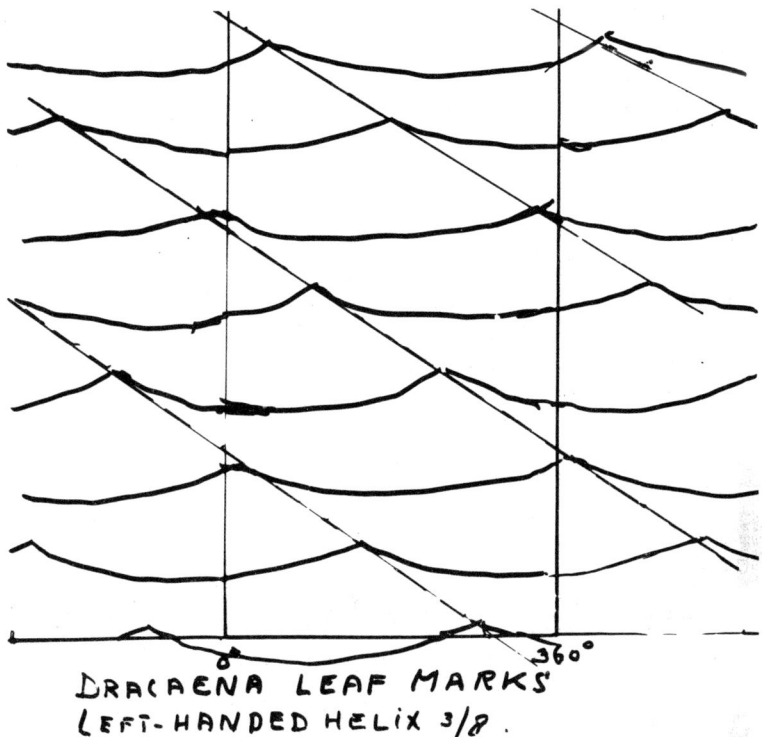

**Figure 22.** Dracaena leaf marks rolled out.

its bare trunk the marks of its shed leaves are still clearly visible. I wrapped a piece of transparent plastic around it and copied the leaf marks. The leaves apparently were arranged along a helix; after five turns, the ninth leaf is approximately above the first, the tenth above the second, etc. So the leaves repeat after 5/8 of a turn, that is, over an angle of 225°. — Note that 5 and 8 are two subsequent members of the Fibonacci series 1, 1, 2, 3, 5, 8, etc., where each member is the sum of the two preceding ones! This series has indeed been found to occur often in the arrangement of leaves in various plants.

I repeated the measurement, and collected the results side by side on one sheet (figure 22). The traces now form a twodimensional periodic pattern. The helix has become a translation. And by choosing an appropriate repeat direction as a second independent translation we can construct a *net* of identical points all over the pattern. We can do the same in the patterns of tiled floors and brick walls

48  *Structures of Matter and Patterns in Science*

which were already mentioned as examples of two-dimensional periodicity.

If there is mirror symmetry in such a pattern, then in any case this net of identical points must be left unchanged by the operation of the mirror. This implies that there is always a translation *in* the mirror and another perpendicular to it. This you can check in many bathroom floors and in brick walls which do have mirror symmetry.

Does rotational symmetry also put restrictions on a net of translation-equivalent points? It is easily seen that twofold rotation symmetry can occur in any two-dimensional net: it just changes the sense of direction of *all* translations. The two figures show examples of twofold symmetry in quite arbitrary nets. M.C. Escher's lizards (figure 23) are in a truly two-dimensional pattern. He made them black and white for the sake of contrast. The molecules of the organic compound (figure 24) are similar in shape to the lizards; here the black and white molecules lie in different planes in space; within a plane the arrangement tail-to-tail and head-to-head is re-

**Figure 23.** Lizards. M.C. Escher. Reprinted with permission of the Escher Foundation, Haags Gemeentemuseum, The Hague.

markably similar to that of Escher's lizards! More-than-twofold rotation symmetry does put restrictions on the translation pattern. Figure 25, a layer in the crystal structure of $H_3O^+ \cdot Cl^-$, shows that threefold rotation symmetry implies a net of equilateral triangles. Note that such a net can also accommodate mirror symmetry.

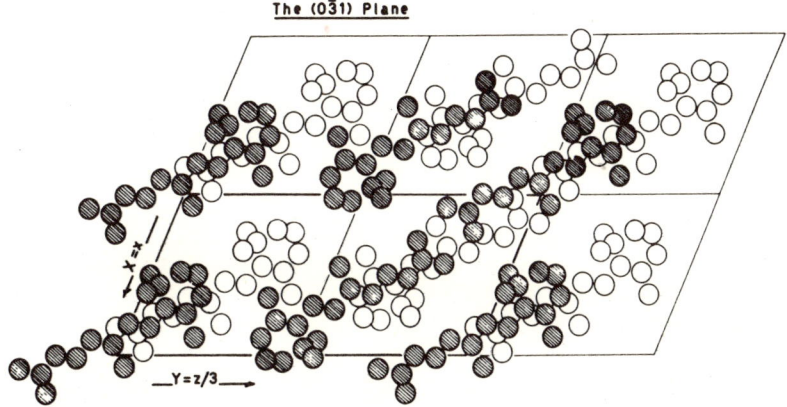

**Figure 24.** The crystal structure of an organic compound (after E.L. Eichhorn).

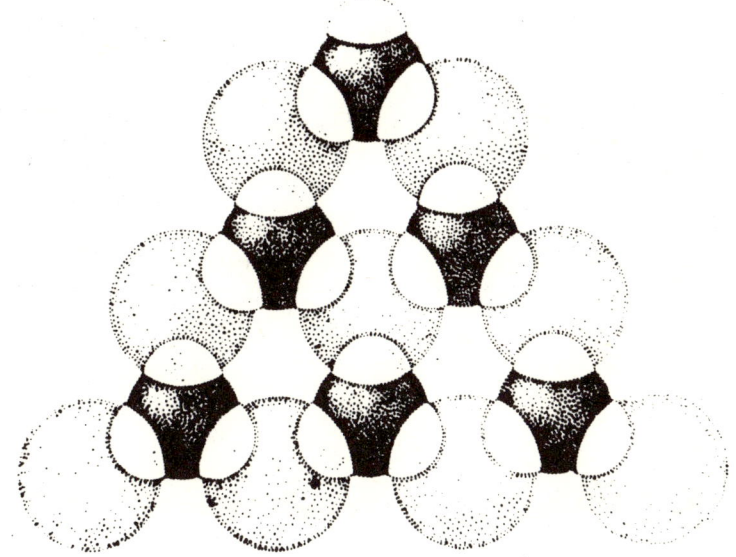

**Figure 25.** A layer in the crystal structure of $H_3O^+ Cl^-$.

The brick walls of the 10th century Samaid Ismail Mausoleum in Bukhara (figure 26) show a pattern with fourfold symmetry — which is also wrapped around the cylindrical columns at the edges. A very similar pattern is found in the crystal of ordinary salt, where sodium and chlorine ions take the place of the motifs of three horizontal and three vertical bricks respectively. The pattern is built on a *square* net, in accordance with the fourfold symmetry.

Sixfold symmetry is presented by a layer of carbon atoms in the mineral graphite, arranged in the well-known honeycomb pattern. By trial you will find that not all the corners of these interlocking hexagons are equivalent by translation. The points that are translation-equivalent, e.g. the centers of every hexagon, again form a net of equilateral triangles. Such nets can thus adapt both threefold and sixfold symmetry. Figure 27 shows an attempt to fill the plan periodically with adjoining five-petaled flowers. You see that it

**Figure 26.** Samaid Ismail Mausoleum, Bukhara. From Milos Hrbas and E. Knobloch, *The Art of Central Asia,* Paul Hamlyn, London.

*Symmetry in Three Kingdoms* 51

simply does not work! Figure 28 shows atoms of uranium and oxygen trying to solve the same problem and arriving at practically the same compromise: a periodic pattern alright, but it does not possess fivefold symmetry. It can indeed be proved mathematically that plane periodic designs only allow 2-, 3-, 4-, or 6-fold rotation.

The same holds for three-dimensionally periodic structures, for example crystals, which can be thought to consist of subsequent periodic layers, as seen in the above examples, figures 24, 25, and 28. Of course, if the whole crystal has to retain the symmetry of the layers, these have to be stacked in such a way that, again, the pattern of translations from one layer to any other is unchanged by the mirror or rotation symmetry. This requires that there is always a translation perpendicular to the layers.

We now come to an important point: I pointed out in figure 15 that the outer faces of that small crystal are, on the molecular scale, densely populated *nets*. This holds for all crystals. If now the atomic

**Figure 27.** Flowers, M.C. Escher. Reprinted with permission of the Escher Foundation, Haags Gemeentemuseum, The Hague.

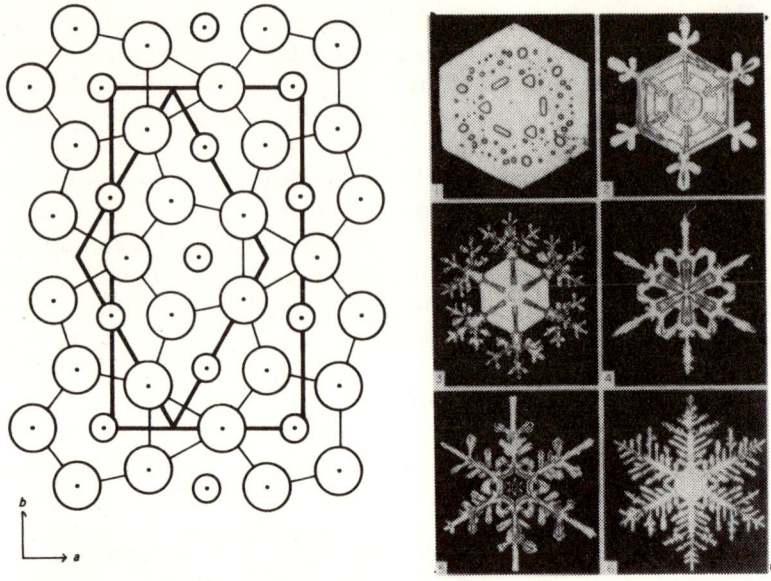

**Figure 28.** Atoms of uranium and oxygen (after B.O. Loopstra).
**Figure 29.** Snow crystals. Photograph by W.A. Bentley.

arrangement of the crystal has a certain symmetry, for example a set of parallel mirror planes, then any face tilted to this mirror must have a completely equivalent face in the mirrored position. So the *macroscopic symmetry* of a crystal, that is, the symmetry of its boundary faces, does tell us something about the *symmetry of the atomic arrangement* in the crystal. The well-known cubes of rock salt (figure 14) betray the regular arrangement of their building blocks, sodium and chlorine ions, on atomic scale. The snow crystals of figure 29, in the same way, indicate the threefold — nearly sixfold — arrangement of the water molecules that constitute them.

On the other hand, some kinds of crystals have themselves no mirror symmetry, but you can find others of the same constitution that are the mirror image of the first. Pasteur concluded long ago that the constituent molecules of such pairs of so-called enantiomorphs must lack mirror symmetry themselves, but one sort of crystals would be composed of "left-hand" molecules, the other of "right-hand" molecules. In the case of quartz (figure 30) this has been proved to be the case. In the atomic arrangement of quartz there are screw-like strings of atoms about threefold screw-axes. In

## Symmetry in Three Kingdoms 53

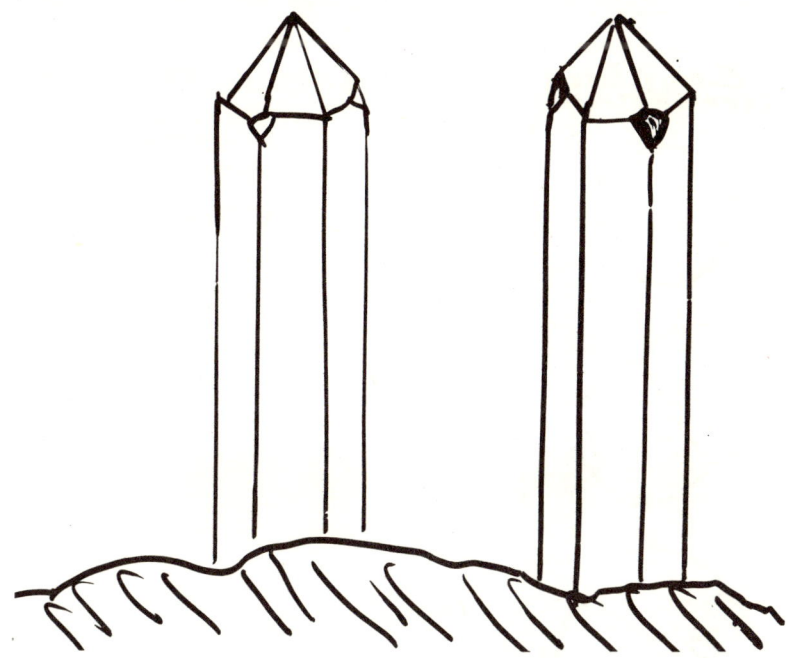

**Figure 30.** Left- and right-handed quartz crystals.

some crystals these are all left-handed screws, in the mirror-image crystals the screws are all right-handed. So again, crystal *shape* may indicate *lack of symmetry* of its atomic arrangement.

Does our own mirror symmetry reflect symmetry of our inner organization? Up to a point, yes: Our skeleton, our senses, lungs, kidneys are bilateral symmetrical; our heart and our stomach are not. But how about our structure on the atomic scale? Our most vitally important molecules, enzymes, chromosomes, etc. are all intrinsically asymmetric: they have no mirror symmetry at all. The same is the case throughout the animal and vegetable worlds. How is it that such asymmetrical molecules tend to build up living organisms with, in many cases, "bilateral," or mirror, external symmetry? One could suppose that such symmetry might already develop starting from the first apparently symmetrical cleavage of the fertilized egg cell. It is true that both daughter cells carry the same genes, that is the blueprint for final development; but a cell contains more than just its chromosomes. Embryological investigation shows

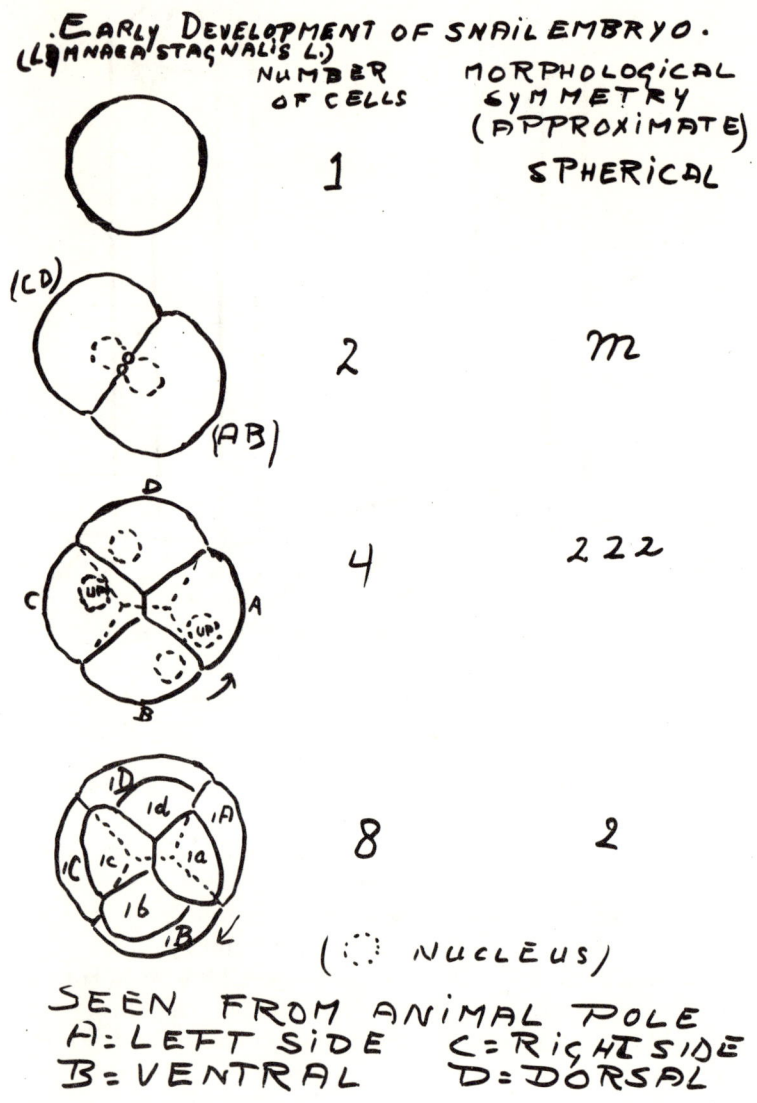

**Figure 31.** The development of the pond snail embryo (after N.H. Verdonk, Doctors Thesis, Utrecht, 1965).

SNAIL EMBRYO, 33 CELLS.
10 HOURS AFTER 1ST CLEAVAGE.

SEEN FROM
VEGETATIVE
POLE

SYMME-
TRY
→ m

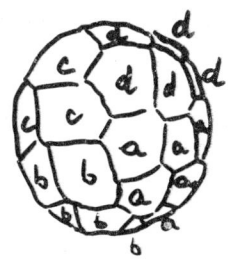

SEEN FROM
ANIMAL
POLE

∼ 2

ANIMAL POLE. 13 hRS AFTER 1ST CLEAVAGE
NUMBER OF CELLS > 50.

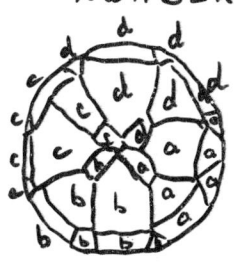

∼ 4 mm

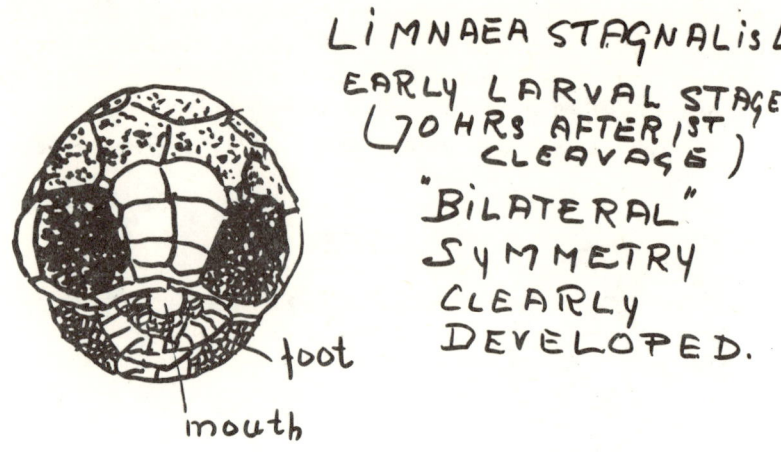

that the two first-formed daughter cells, although they may look very similar in shape under the microscope, mostly develop differently. For example, in some algae one daughter sprouts a sort of root to anchor the future plant, while the other daughter and her offspring grow towards the light. But these algae are odd-shaped creatures anyhow. How about a more symmetrical organism? Figure 31a, b and c show sketches of the early development of the embryo of the pond snail, with the morphological symmetry at each stage. Apparent mirror symmetry develops only after many cleavages.

Then how and why does symmetry of living organisms develop? Let us start again with the unicellular radiolaria. They float about willynilly. Their skeletons are either bell-shaped or isometric. Their spikes support them and give some guidance to their protoplasm shooting out in thin threads in all directions.

Sedentary sea-anemones also need many "arms" to catch their prey. Their high rotational symmetry helps in the business.

Bilateral animals move purposefully ahead. Mirror symmetry is useful to them. They develop sense organs in front, so they can avoid collisions and detect prey or enemy.

Some flowers spend their pollen and/or seeds freely: they have multirotational symmetry. Others hide their sexual organs shyly inside. They must have bilateral symmetry to fit the bilateral symmetry of the fertilizing insects. So symmetry among living organisms appears to be the result of evolution, of mutual adaptation of way of living and environment. That is: *their symmetry is functional,* and gives no information about the arrangement of their constituent particles on the atomic scale. The same holds for contraptions made by animals: birds nests, spider webs, and other traps for catching prey, honeycombs, and all things made by men for practical purposes.

*Conclusion.* Macroscopic symmetry in the *mineral* world, in particular in crystals, is related to, and an indication of, the symmetry of the atomic arrangement in the crystal. Symmetry of *living organisms* and of their products is *functional* it serves the purpose of life. The *properties* of symmetry, the various kinds of symmetry elements and the laws governing and restricting their combinations, are *universal:* symmetry is a property of *space* as such, and its laws can be derived mathematically. In the above, I have avoided all mathematics, but I hope that the examples I described have conveyed an idea of the universal character of symmetry.

# SCULPTURAL MODELS, MODULAR SCULPTURES

Arthur Loeb

*In the beginning, there were the tetrahedron and the octahedron. The beginning, that is, of my friendship with Dorothy Wrinch. I met Dorothy as a result of these very basic shapes; that is not surprising because we both were interested in the basic building blocks that make matter, and life, around us. These models led me into a great many adventures, wonderful ones, and I met a great many marvelous people that way.*

They led to an article in the series that Gyorgy Kepes published, that fine series on shapes, called VISION AND VALUE, in which there was one particular book dealing with symmetry, the module, proportion, rhythm.[1] And I believe it was through that particular article that Marjorie Senechal got in touch with me. There were a number of years when you at Smith were having classes on Armistice Day, and we in Cambridge were not; I am sure for many of you this was very aggravating, but I found it very pleasant, because that was the occasion on which Professor Senechal often asked me to come up and talk to her class.

The first time this happened, I asked Marjorie whether she knew a Professor Dorothy Wrinch who had written me about my Moduledra crystal modules. She sounded like a very interesting person. I knew about her, I knew her publications, her work, and I looked forward to meeting her. That is how I did meet Dorothy: as you know, in November, Armistice Day is followed by Thanksgiving, and we discovered very soon that Pamela Wrinch Schenkman, Dorothy's daughter, happened to be living half a block away from us in Cambridge, and Dorothy was planning to spend Thanksgiving holiday with her daughter and son-in-law. So two weeks after our meeting, we were sitting in our music room playing 4-hand piano: this started off a very harmonious relationship. I think it is very significant that our meeting came about through the medium of art, primarily, in this particular case, through the communication skills of Gyorgy Kepes with his volumes that have produced so much synergetic interaction between people.

A great deal is being said nowadays about the differences and the similarities in the creative process for artists and scientists. I myself don't understand the creative process so I am not going to speculate about this very, very difficult question. I would like to say, though, that there are great differences in some of the functions of the scientist and the artist. The scientist is not always a great communicator, although at this particular symposium I think we are trying our very best. The artist is primarily a communicator. An artist without an audience, without somebody to address, is essentially meaningless. So, I feel the collaboration of the scientist and the artist, whether it is between several persons or whether it is within one particular person, is indeed very fortunate. We saw that happen last night in Caroline MacGillavry's lecture. The collaboration between the scientist and the artist is of extreme importance, and I would like to talk about this process a little bit, particularly the visual thinking that was so important to Dorothy.

We tend very much to be trained and to be selected on a verbal basis by the tests we take. A great deal of the work that we do in solid state chemistry or crystallography is not scalar — that is to say, it need not deal with strings of symbols along a line, but with the symbols in space — in three dimensions. I feel that it is enormously important in dealing with spatial complexity to realize that these are not strings of symbols on a scalar line, but that they are in what we call a vector space.

Let us consider some models. Figure 1 is a conventional model of spinel. Spinel is a mineral, but it is also a man-made structure. The man-made structures are ceramic materials very much like the mineral spinel; they have exactly the same configuration of atoms, although the kinds of atoms may be different. Some years ago, we were looking at this particular model when we were designing the first memory cores made out of magnetic ceramic for the Whirlwind computer. This is important background information because in working with computers, I was very much conditioned to think in terms of binary digits of information, of processing information in such a way that it requires a minimum number of digits — binary digits — to communicate, to store, and to retrieve. I realized in dealing with this particular model that the amount of information was formidable, so formidable that whenever we developed an insight into it we would quickly lose it again. I remember how on a

## Sculptural Models, Modular Sculptures 61

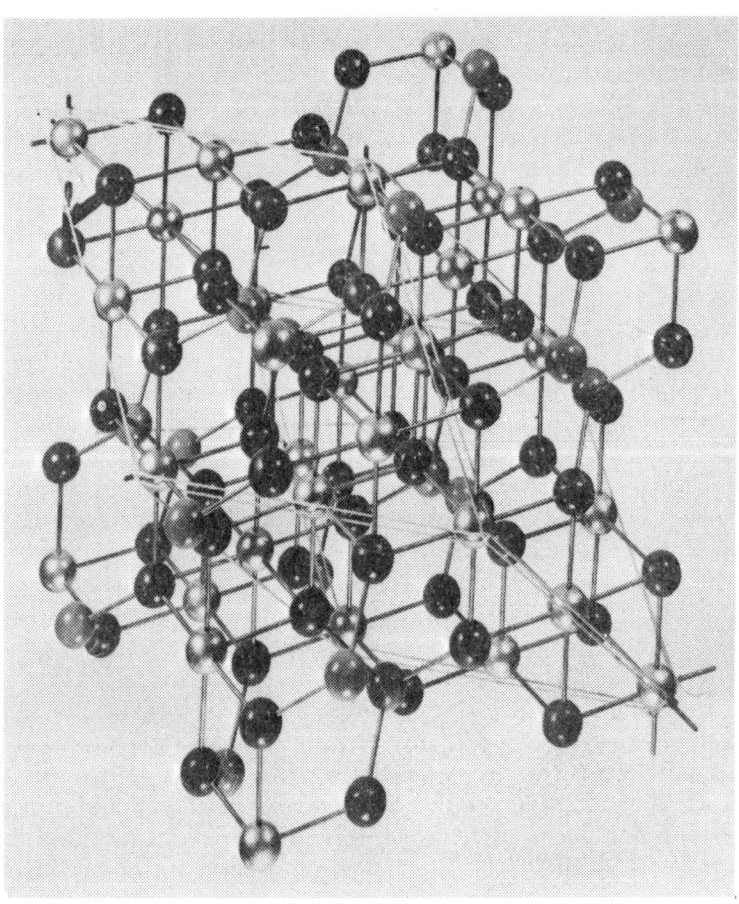

Tuesday afternoon we spent hours, and by Wednesday morning we had to start thinking all over again, because we had no particular way of storing, or recording, our insights. We were illiterates, spatial illiterates. We were people who have no way of writing down the conclusions they have come to after complex discussion and consequently there was no way of recalling it the next day.

So I decided to analyze this problem in terms of that information. I realized that the metals were always either octahedrally or tetrahedrally surrounded. The oxygens around them were then at the corners, 6 of them at the corners of an octahedron, or 4 of them at the corners of a tetrahedron. The question then was which metal ion was to go into which kind of situation to find the proper environment. I decided that the best way to look at it would be to think of what would happen when these particular ions were by themselves in the company of oxygen. If we could see what their most preferred arrangement would be in such an environment, then how would they compete to create a crystal that has the correct configuration? I realized that the tetrahedron and octahedron were enormously important.

It was some time after that, on a very hot and muggy summer afternoon, that I decided at 2 o'clock to take a bit of a siesta. I lay down, and I closed my eyes and I asked myself, "What do I know about the octahedron? What do I know when I don't have this shape in front of me, but imagine putting it on the table, and ask myself how many edges are vertical, how many edges are horizontal, how many edges are at what kind of other angle?" I realized that that sounds like a very simple problem and you can see it solved when you have the octahedron in front of you; when you haven't got an octahedron, it is not so simple, because we are not used to thinking in terms of edges that make triangles, that go at peculiar angles, where you have no verticals at all. I concluded that that kind of information should really be prefabricated. If we can prefabricate that kind of information and build these modules, then we should be able to fit those together in a great many ways. The amount of information that is necessary to create such a structure, once you have the complex information prefabricated in your blocks, should thus be considerably reduced. Accordingly, I designed four modular blocks, two tetrahedral, two octahedral, to serve as modules from which the spinel structure and many other mineral structures can be modeled by assembling them in different permutations and combinations. I had in mind making a good attractive design, and that's been very helpful, because in that way I attracted the interest of artists and consequently got into so many of these adventures.

One of my colleagues, David Shoemaker at M.I.T., borrowed some of these Moduledra. One day I had to see him about some other matter, but he said, "Oh, by the way, I was working on zinc cyanide, and I used your Moduledra to make a model." I looked at it for maybe 20 or 30 seconds, and I replied, "How nice." And then we went on with our business. Six weeks later we were together at a meeting of the American Crystallographic Association. At that time he was giving a talk about zinc cyanide and I was giving a talk about the use of the modules. I asked whether he would like me to build

## Sculptural Models, Modular Sculptures 63

the model for his talk of the zinc cyanides. He replied, "Well, it takes me quite a long time. I don't think there is time to do it." However, in about 2 or 3 minutes — I had it built, and he looked at me and said, "It's absolutely correct, but how could you remember? You only looked at the model for 20 or 30 seconds six weeks ago." The interesting thing is that all I needed was half a minute, to see that 12 binary digits of information sufficed for putting these together. I had no trouble remembering those 12 binary digits, so I just built it up again. The information channel, using binary digits, was not really very much overloaded by that amount of information. It was the way in which we analyzed the problem and the way in which I found the complex information prefabricated which enabled me to reconstruct the entire model very easily. This experience gave me a great deal to think about. It influenced a lot of my work after that, because I realized that there is indeed a special way in which we look at space that is tremendously important, and if we minimize the amount of information in communicating, then we solve a really very important problem in the storage, communication, and retrieval of spatial concepts and patterns.

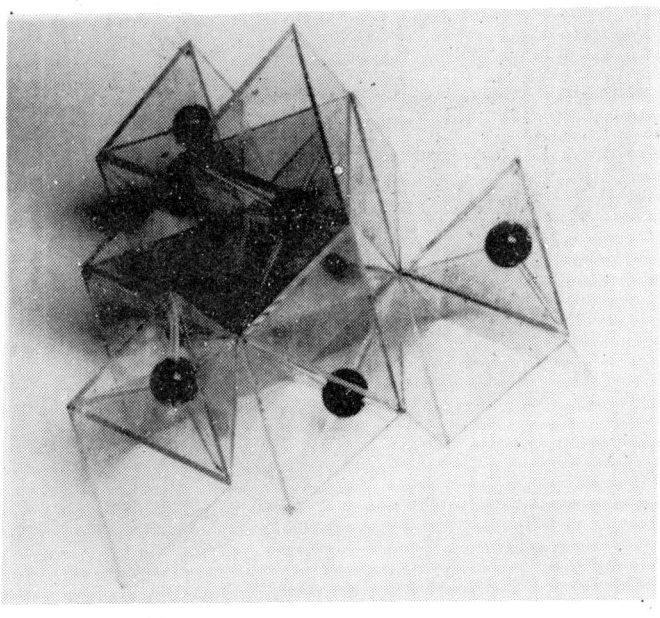

Figure 2 shows my model of spinel. You see that the complexity is much reduced. This is just as good a working model as the one in Figure 1. It gives all the information you need and it's easy to remember.

Figure 3 gives a so-called structure of structures. It tells in a list, the models of minerals which I can build up, using these building blocks in various permutations and combinations. But it does more than that, because it tells us a relationship between structures, and this is very important when you think about structures and patterns. Perhaps at this time, I should define my terms. An array is an assembly of things; a pattern is an ordered array in which the individual things have a defined relationship to all the others; structure is the set of relationships between them. It is a somewhat specialized way of thinking of structure; we usually think of a structure as the thing itself. But basically when we talk about the structure of an organization, or the structure of a crystal, what we really are saying is that we are interested in the relationship between the individuals in the organization, or between the atoms and ions or the molecules in a crystal. When I talk about a structure of structures, I am really saying that each entity has internal complexity. But there are also, *between* all these entities, very definite relationships, that help us in transforming one into the other. There are several examples at this symposium of a structure of structures. One of them, for instance, is David Harker's colored symmetry models (which are in the exhibition in McConnel). In talking about chirality, what he is basically saying is: we have several structures here, and internally they have identical symmetry. Each color repeats with the same frequency and at the same distance and configuration within any model. But a given permutation of these colors turns one model into the other. So there is a very definite relationship between these different structures which then defines a structure of structures.

In 1965, Duncan Stuart of the School of Design, University of North Carolina, Raleigh, and I were invited to create an exhibition "Symmetry and Transformations" at the Carpenter Center for the Visual Arts at Harvard. We had never met, and for a half year we worked independently, Duncan as a designer, a former student of Buckminster Fuller, I as a geometrist, applied mathematician, chemist, whatever. They only introduced us when we brought up our work after a half year of preparation, approximately a half year before the opening of the exhibition. Interestingly enough, the designer came with many verbal panels, and I came only with visual

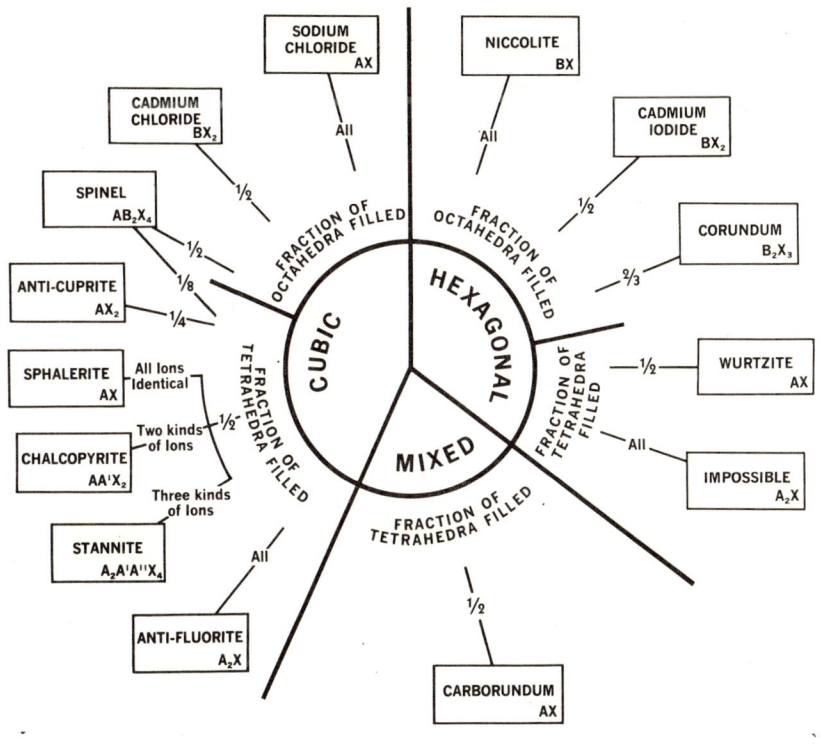

panels. This was amusing, significant, and interesting. The show was put together back to back and at various places you could peek from one side to the other and find the concordances.

I think that you are familiar with some of the interactions of scientists and artists. For example, we all know the work of M.C. Escher. I see the significance of Escher very much as that of a communicator. Escher had an exhibition at an international mathematical congress in Amsterdam in the early '50's. He was welcomed by the scientific community as a communicator. We know what he did with some of Penrose and Coxeter's mathematical indeas in three dimensions, and we are all familiar with Caroline MacGillavry's book on the symmetry aspects of Escher's art.

I myself was, in the early '60's, very much involved with Kepes, and with Escher and Buckminster Fuller. I was invited here to give a talk by the Student Lecture Committee in the early '70's on Structure in Art and Science. At about the same time I wrote an article for the journal *Leonardo*,[2] which deals very much with the interaction between art and science. I want to show you a tryptich that makes up a composition, which I designed for the article in Leonardo. This triptych is called *Gebonden Vrijheid*, which is Dutch. It comes from my school motto, from the Barleus Gymnasium, which was *Disciplina Vitae Scipio*, discipline is the staff of life. What I wanted to point out to the artists and the scientists in this article was the fact that you can begin with an enormously disciplined design. I had a very rigorous rule. I started with two triangles (figure 4) which are each others' mirror images. You see there is a mirror that reflects one triangle into the other one, and vice versa. There is an overlap, so this is the kind of mirror through which you pass part way, like Alice in Wonderland. I also required each of these triangles to have a 3-fold center of rotational symmetry at its center. That is to say — each triangle should be surrounded identically by 3 segments of space. But each new triangle has in turn to be surrounded by 3, so we would go on indefinitely, creating this pattern with just the rule that there had to be pairs of triangles related to each other by mirror symmetry and each triangle had itself to have, at its center, a center for 3-fold rotational symmetry.

The next thing that I did was deliberately to lose sight of the original generating triangle. I looked instead at the interspaces and the overlap spaces, which were created by the triangle and the generating rule, and also at these rather irregular interstitial hexagons that have 3-fold rotational symmetry. I looked at those spaces, and you see what happens in the second member of the triptych. The triangles are still there, but they are very much obliterated by the emphasis on the colored spaces. Now the question was, what do you do after that? I began to think about the tensions created, this triangle still calling for attention, and the other spaces that were just discovered almost taking that attention away. And so, in the third one, I was quite free. Some people like what they call "the random one" best of all. However, there is nothing random about this, but there are artistic decisions that are not bound by the original rules. The original rules had laid the framework, had used the triangles, had created the tensions; the hardest part was to make a design in which all of these components, visual components, compete for attention and cause the tensions to balance out. So you see, the three seen together show you the disciplined freedom that the artist can have, by starting with a fairly rigorous rule that defines the attractive spaces in a way which I think he would not just freely devise. People who say, "Nature is an artist" and "snowflakes are a work of art,"

## Sculptural Models, Modular Sculptures

express an appreciation of Nature. We all admire Nature, but Nature is not an artist. It's by what we see, by how we arrange things, that we are artists, because art always involves a human decision, and you cannot say Nature is an artist, unless you are so anthropomorphic about Nature that you think of Nature as a goddess.

The Institute for Contemporary Art in London asked me to give a lecture which we called "Putting Multiples Together", and to make a visual design that would incorporate putting multiples together. So (figure 5) I took the word Multiples — M U L T I P L E S — and, realizing that the M has a vertical line of mirror symmetry, and that the letter S has, at its center, a center of two-fold rotational symmetry, I took that mirror, I reflected MULTIPLES back, I took a center at S, and flipped the thing over 180 degrees. So we now have MULTIPLES upside down, because it was flipped 180 degrees around a point. I created a framework which indeed was created by MULTIPLES put together, and nothing else. There isn't a black line in it that wasn't formed by the word MULTIPLES! I then again used my artistic decision, my freedom, after this discipline, to color it; this is what is at the Institute in London now: PUTTING MULTIPLES TOGETHER.

In a very much more conscious and definite way, I worked with the Committee on Programmed Instruction at Harvard, which was part of Skinner's laboratory, on programming this information. Again it was an exhibition which had brought these models to the

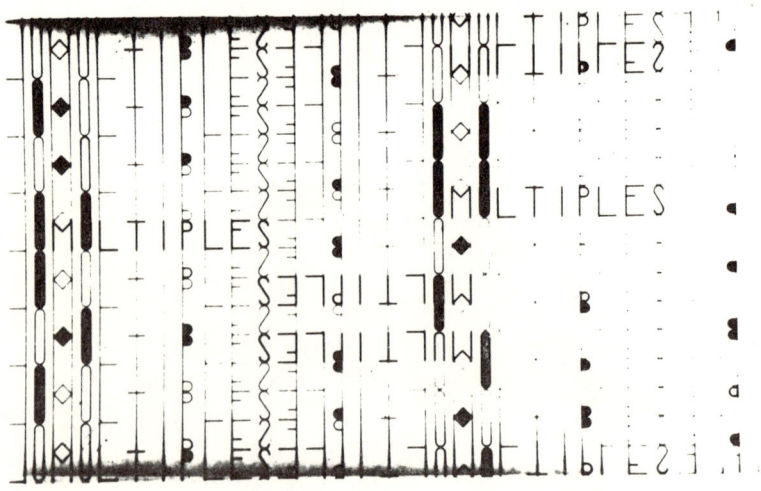

attention of the staff of the Committee. They had been asked to write a program of self-instruction involving learning to deal with visual spatial complexities, particularly in 3 dimensions but also in 2 dimensions. They asked me to design a teaching machine that incorporated models — doing whatever I liked electronically — to see how we can teach people to think spatially. Well, Eric Haughton of the Committee Staff and I felt that we should not make complex machines, because it is difficult to wean people away from them. They will learn when they are tied in to a computer or any other kind of teaching machine, but how do they really learn to visualize and think 3-dimensionally? It is a difficult question — to what extent you can teach this to everybody — some people are more gifted than others. I myself see very poorly in 3 dimensions. I cannot fuse, I cannot look at a stereogram, but I can think and I think that the same thing is true of Buckminster Fuller and Duncan Stewart. I think we have had to compensate with our reasoning for a lot that we can't see directly. Some people are more gifted that way than others, but nevertheless, you want to teach them a certain amount of the basic science of configurations, even if they are not particularly interested in 3-dimensional problems. So I immersed myself in learning just what programmed instruction was, and I came up with two conclusions. Going through some practice programs, I was impressed with the fact that you have to analyze the tasks into very small units, which can be sequenced. That, I think, is very, very good. I was used to that in programming for computers anyway. I also liked the fact that you give a certain degree of reinforcement by telling a person right away in these small tasks whether he or she has made the right decision or reached the right conclusion, or come up with the right answer. But I was disturbed by the fact that all these programs were so verbal; it seemed to me that one could develop a knack of coming up with the right word for the answer without really understanding the concept, and as a result, one would tend to forget it very soon. So, Eric Haughton and I decided to do some experiments in nonverbal programming which means using no words at all. Figure 6 shows a number of frames from a program sequence.[3,4] You have to realize that these are not always consecutive ones. The numbers in the left-hand corner show the location in the program sequence; I chose eight with lots of other frames in between. The sequence was written for a very simple and ingenious learning machine. Each slide was projected onto the machine panel; there was a way of touching any one of 5 sections of the panel; so that if you touched the correct one,

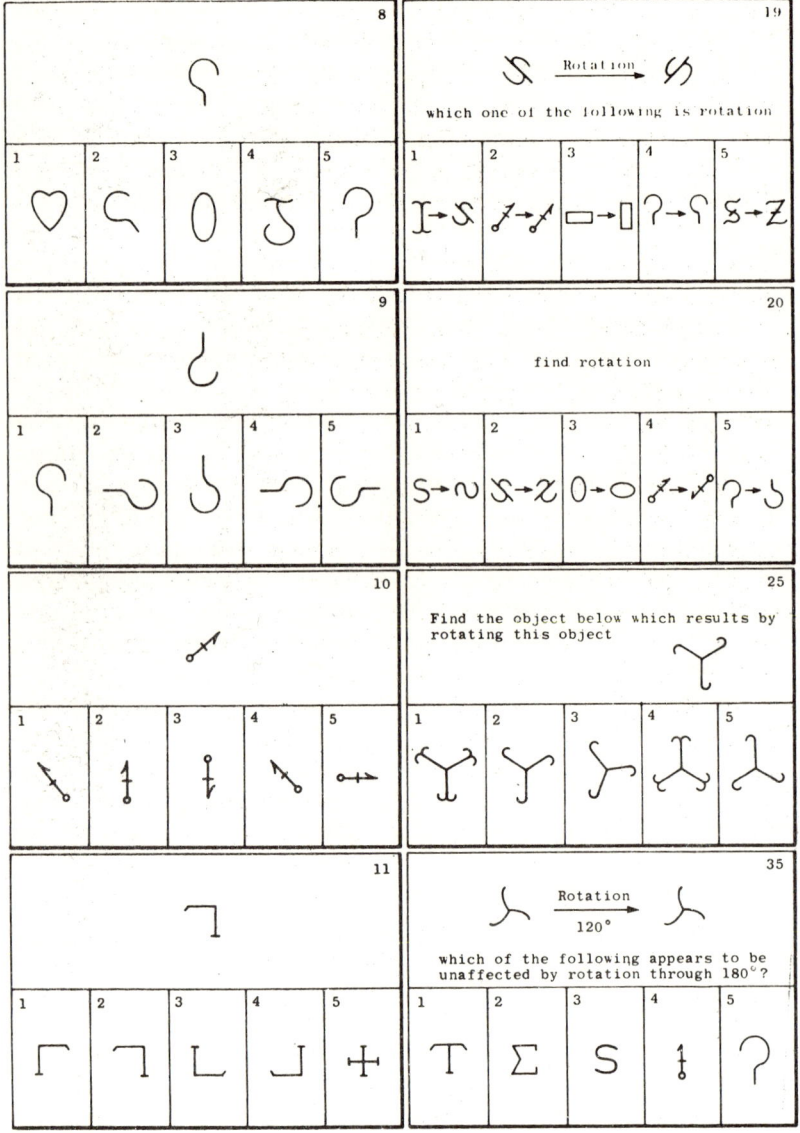

**Figure 6.** Examples of the multiple choice, nonverbal symmetry program. These frames have the sample or material to be responded to in the upper portions, with the alternatives in the five choice panels. Appropriate responses are 8-2, 9-2, 10-1, 11-2; 19-3, 20-3, 25-5; 35-3, 38-3, 39-5; 48-3; 63-2; 185-1; 196-2.

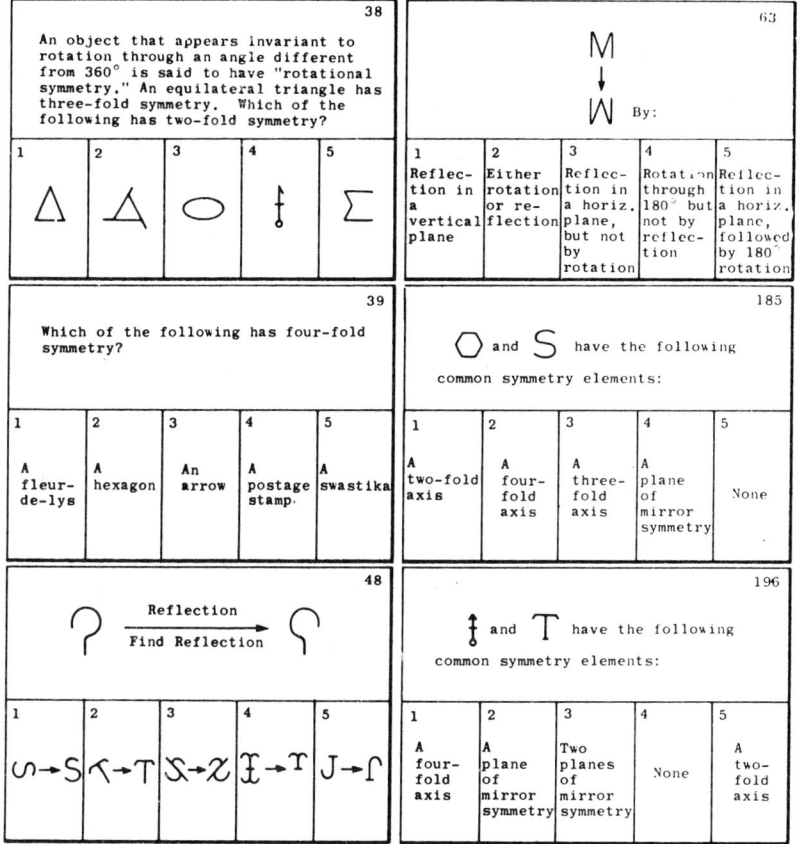

**Figure 6,** continued (see legend on facing page).

the machine would advance to the next slide. We started very simply, without words, by giving the identical object at the top and down below, so that people would get the notion that they had to do something when they saw identical objects. We had really given them no verbal instructions; that is important to remember. We just started them that way. At number 9, they would have to compare which panel matched the ordinary question mark. That is the panel that they would have to touch. If you touched another panel, you would have realized that it does not contain the appropriate image, having by now learned to recognize that. In the next frame we have essentially the same thing. You see a question of discrimination between an object and its mirror image. Next we have come to the first words. Here we say, "Find rotation." And now we have a transformation, and they have to discriminate between a rotation and a reflection. So first they learned discriminations; we were very careful not to use words until the concept was demonstrated to be clearly in people's minds. In carrying out these experiments, we took a group at random — for example, some people who had a summer job in an office there — and we gave them a task. We always made certain that we gave them a test before they went through the program, and afterwards we gave them the same test again and compared test scores on the pre-test and on the post-test. We found something very interesting. The task at that time was to take the upper case letters of the alphabet, all of them — A, B, through Z, and to put them into different classes according to whatever symmetry they had. For instance, the A and T both have vertical mirrors, so we wanted them put in the same box. The letter N and the letter S both have only two-fold rotational symmetry so they would together go into a second box. It was remarkable that in the pre-test, of course, they had no idea how to classify these letters, and so they scored very low. After an hour, they did extremely well — most of them infallibly classified these letters. But when we asked what they had learned, they replied, "Why, nothing." We said, "Well, but you were able to do a task afterward that you weren't able to do before." But they felt that there had been no effort involved, so they obviously hadn't learned anything. So we remarked, "But you must have learned something. You were able to do that task afterwards, and not before." The reply was: "Oh we did it intuitively." Now you see, they hadn't learned anything; they did it intuitively, but they didn't have that intuition before! At 12 o'clock they didn't have the intuition, but at one o'clock they did. It gave me considerable insight into what people call intuitive. I think that a lot of what is called intuitive is, in fact, nonverbalized knowledge. You see, we had not given them the words to express what they had learned, so they couldn't tell us. Therefore they thought they had learned nothing. This pre-experiment really gave people a concept without words. They weren't able to express themselves at all, so they said, "We learned nothing." I think that is very significant in this whole notion of visual thinking.

My experiences in designing materials capable of containing binary information and programs for teaching spatial visualization have convinced me that our society needs to learn to cope with spatial complexities in a quantitative manner. This means the development of a nonlinear language, a language that does not consist of linear strings of symbols. I have some indications that such an approach is successful in teaching mathematics to dyslectics. In any case, we need to understand three-dimensionsional spaces (or two, or four-dimensional spaces) on their own terms, not as stacked one-dimensional strings of symbols. Architects need to understand how to optimize personal space, how to create nonintersecting traffic flows (e.g. pedestrian and vehicular) in three-dimensional spaces. Sculptors must learn how to transform spaces, and what is necessary to stabilize spatial structures. Elsewhere[5] I am reporting on our Design Science Studio at Harvard, in which students in the Arts and the Sciences together learn both to express themselves three-dimensionally and to expand their three-dimensional repertoire. Let me conclude here with a photographic essay using one of our sculptural models/modular sculptures which may be used either as a traveling stage set (its modular components open up to give a wide variety of settings, yet pack compactly inside a cubic box), or to illustrate a powerful property of three-dimensional space.

Two-dimensional surfaces may be subdivided into mutually congruent tiles in such a way that half the tiles have one color, the other half a different color. For instance, a plane my be divided into white and black squares, such as on a checkerboard, so that each white square shares edges with four black ones, and vice versa. Or the plane may be divided into green and blue triangles such that each blue triangles shares edges with three green ones, and vice versa. In neither case does *every* tile share an edge with a tile of the same color. Consequently, a continuous region of either color must interrupt the region of the other color. This is a general property of the plane.

In three-dimensional space the situation is quite different. Con-

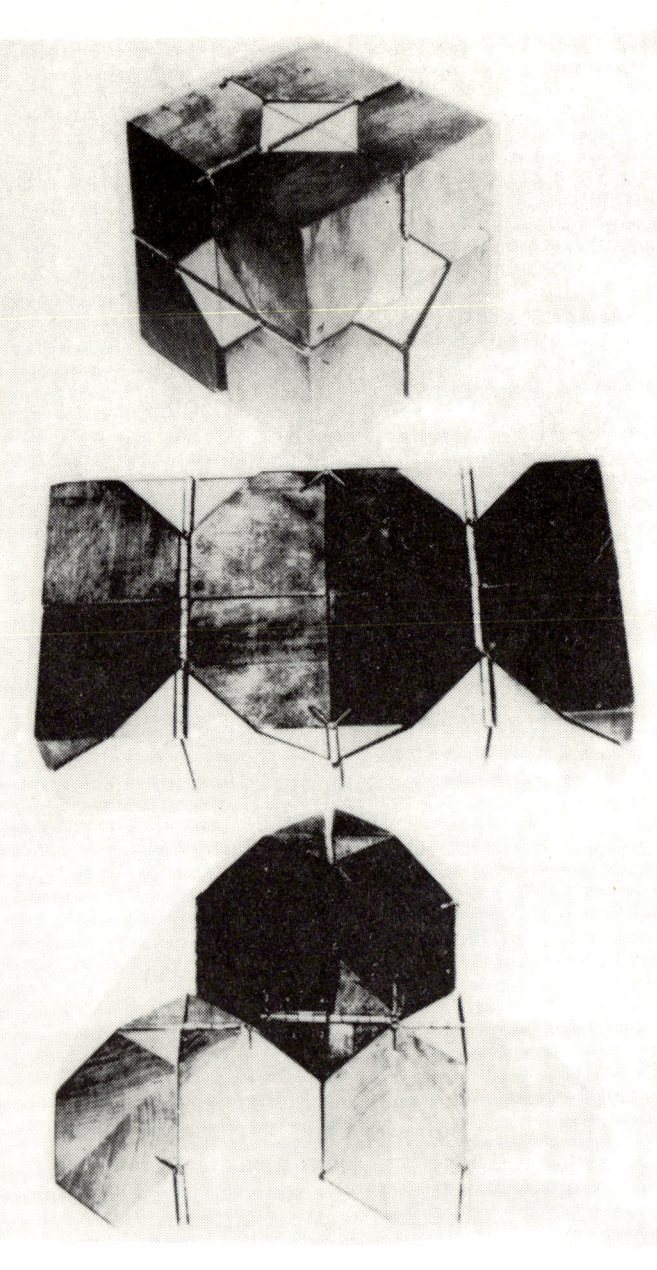

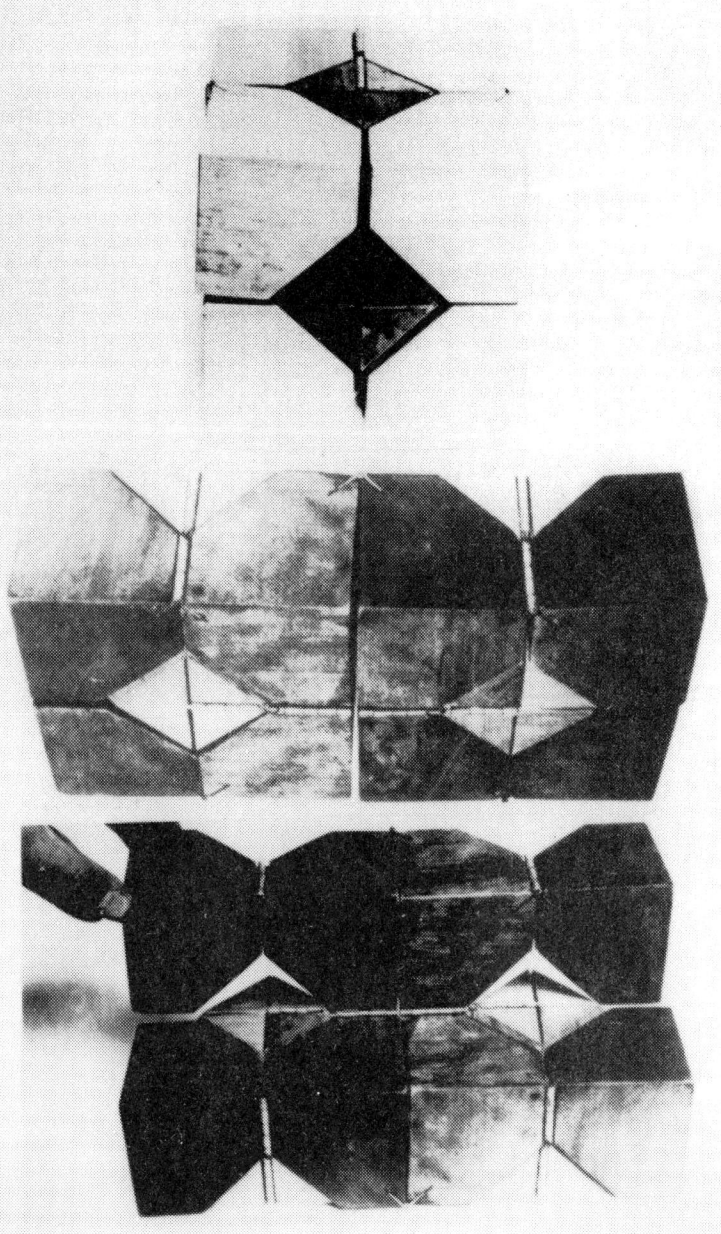

sider, for instance, a truncated octahedron which has eight hexagonal and six square faces. Such a polyhedron together with mutually congruent copies can fill all of space in such a way that, say a black one shares it hexagonal faces with eight white ones, and its six square faces with six other black ones. Each white polyhedron then shares its hexagonal faces with eight black ones, and its square faces with six other white ones. Such stacking divides three-dimensional space into two subspaces, each continuous, which interpenetrate but never intersect each other.

Figures 7 illustrate these properties of the truncated octahedron. First we see the truncated octahedron inside a cube; this configuration is transformed into one in which the light and the dark modules are interchanged. We see thus that the central truncated octahedron is entirely equivalent to the space inside the cube but outside the truncated octahedron. In point of fact, the latter space is made up of eight octants of a truncated octahedron. The vertices of the cube are thus entirely equivalent to the center of the cube: when the center of the cube is also the center of a light-colored truncated octahedron, the cube vertices are the centers of eight dark-colored adjacent truncated octahedra. The transformation interchanges the location of the dark- and light-colored polyhedra. The other illustrations of Fig. 7 show the modules in various sculptural manifestations, stressing the equivalence of the truncated octahedron to the space left in the cube after the truncated octahedron has been removed.

# DECIPHERING PROTEIN DESIGNS*

Carolyn Cohen

## PREFACE

*I have the privilege of talking with you this evening about a subject that was of the deepest interest to Dorothy Wrinch: that of the structure of proteins. As you probably know, Dorothy Wrinch began her creative work as a mathematician and philosopher, and in the 1930s turned her attention to Biology. She became a member of a small remarkable group at Cambridge University — the Theoretical Biology Club. Joseph Needham's classic book,* Order and Life *(1936), was, to a large extent, generated from their discussions. It is dedicated to members of that club: among them are J.D. Bernal, J.H. Woodger, C.H. Waddington; Dorothy Needham and Dorothy Wrinch. The full story of this group is not yet known, but they were primarily concerned with the analysis of biological form – both its philosophical and physical basis. And their common belief was in the vital importance of proteins as the key structures in Biology. Dorothy Wrinch's life's work centered on this problem, and she influenced many, including Joseph Needham in England and, in America, Ross Harrison, the great embryologist at Yale, and Irving Langmuir, the physical chemist. I believe that her influence has been vastly underestimated.*

*As with other speakers here, my own first encounter with Dorothy Wrinch occurred at Woods Hole. It was a Friday evening seminar in the summer of 1949 when she showed her cyclol models (frontispiece). The beauty of these structures and the clear conception she had of the importance of proteins led me to appreciate, for the first time, the significance of this problem. I present this paper as a tribute to a gifted and remarkable woman.*

---

*This work was supported by grants from the National Science Foundation (PCM 76-10558), from the National Institutes of Health (AM 17346), and from the Muscular Dystrophy Association.

# INTRODUCTION

Riddles are the stuff of science: seeing the riddle plainly is one part of the puzzle; solving it the other. But riddles characterize also the stages in our culture. In his book, *Homo Ludens*[1], Johan Huizinga has described how "the riddle... was originally a sacred game." For archaic man, a profound connection was established between playing and knowing. The cosmic order was celebrated in rites and contests, such as the Olympian games, in which physical strength or cunning was put to the test. The outcome might decide the life or death of the player. Oedipus became king of Thebes by solving the riddle of the Sphinx. He thereby escaped death and restored life to the city. The heroes Heracles and Odysseus had to overcome superhuman obstacles in their trials. Thus, the essence of classical Greek tragedy and myths is man's attempts to solve riddles of a divine order — the clues to which the ordinary mortal cannot see. From these beginnings, the dialogue took form and was brought to an art in the verbal play of the Sophists. As Huizinga has shown, Greek philosophy emerged from these riddle contests.

In a very direct way, we can also trace the emergence of natural science from philosophy. Here, again, the riddle has remained a touchstone. Better said, we can see a new use of the dialogue in the 16th and 17th centuries, when science emerged from Scholasticism. The dialogue was no longer an arid ritual taking place between man and an established dogma but — with the use of the experimental method — it became a basic dialogue between man and nature itself. A contest of a sort was set up, where the scientist, by the design of an experiment, forced nature to answer "Yes" or "No" to a question. Let us, as an example, take the discovery of the circulation of the blood in the human body. Up to the time of William Harvey, the early 17th century, it was thought that blood moved in the manner described some fourteen centuries earlier by the celebrated Greek physician Galen. Galen decreed that the blood, which was different in the right and left portions of the heart, moved with a kind of ebb and flow, each in its own domain. As George Sarton has pointed out,[2] Galen had to assume that the blood passed through some invisible pores in the solid wall dividing the right and left portions of the heart. And even the great Leonardo da Vinci, who had dissected and examined so many human hearts, was too bound by Galen's dogma to question what he saw. But Harvey was not so bound and made the great discovery, which he described in a marvelous little essay in 1628, *De Motu Cordis*.[3] His experiments were simple and ingenious: involving, for example, no more than a tourniquet about the arm to show that when a vessel is constricted, the direction of blood flow can be determined. Harvey showed that the blood is pumped from the heart and passes by way of the arteries through the body to return by way of the veins to the heart; that is, that the blood circulates in the human body.

Contemporary scientists are part of that experimental tradition. They are natural philosophers, engaged in the agon and play of the riddles of nature. And in the field of the life sciences, I think it is fair to say that no puzzle has been as challenging and no solutions as surprising as those of the structure of proteins.

## WHAT ARE PROTEINS?

The qualities that we associate with life depend directly on the class of molecules called proteins. We are, for the most part, composed of proteins, and we can function only by means of proteins. Jacob has noted that it is often claimed that Harvey furthered the mechanistic view of living things by comparing the heart to a pump and the circulation to a hydraulic system. But this is an inversion of the order of events. "In reality it was because the heart works like a pump that it was amenable to study. It was because circulation can be analyzed in terms of volume, flow and speed that Harvey could perform with blood experiments similar to those which Galileo carried out with stones."[4] Similarly, biological molecules can in some ways be considered as machines and "of all . . . the molecules found in nature, proteins are probably the most varied, the most complex and the largest."[5] The absolute uniqueness of protein molecules in their design long delayed the solution to their structure and function. They are like nothing else on earth!

Now just what are proteins? Their chemistry provides the first clues. In the early 1800s it was recognized that certain biological molecules containing nitrogen had extraordinary properties. Their size could not then be determined precisely, but it was known that they were giant molecules, or macromolecules, and that some of them were associated with remarkable catalytic processes taking place in living things. These substances were named proteins by the great chemist Berzelius. At the very beginning of this century, Emil Fischer showed that proteins were, in fact, polymer chains made up of repeating units called amino acids and held together by strong linkages called "peptide bonds" (figure 1). In order to break these linkages, high temperatures and acid or alkali are needed. The

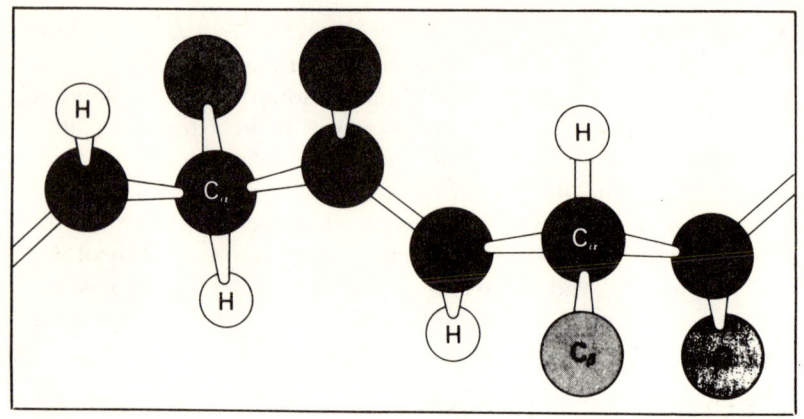

**Figure 1.** Polypeptide linkage. Conformation of a dipeptide unit. The polypeptide chain is almost fully stretched out. Reprinted with permission from Lubert Stryer, *Biochemistry*, Copyright © 1975 by W.H. Freeman and Company.

organic chemists then revealed the surprising fact that all proteins were composed of amino acids belonging to only 20 different types. Each of these subunits has a similar part that forms the links of the chain and thus what may be called the "backbone" of the protein structure. But one amino acid differs from another by virtue of its side chain. These side chains have very different chemical groups and properties: some are hydrocarbons resembling oils and are not soluble in water; others carry an electric charge and are water soluble. Thus, the basic chemical structure of proteins is the polypeptide chain, differing in the composition and ordering of the 20 types of amino acids, and differing in size by containing from several hundred to tens of thousands of amino acid subunits.

This is but one part of the story. We shall see later that the *sequence* of amino acids directly determines the folding of the chain — that is, the three-dimensional structure of the protein — and that, as yet, scientists cannot predict from the sequence how a particular protein will fold. Only the protein chain knows its own "inner logic,"[6] involving weak interactions among the many side chain and backbone atoms. We must emphasize that these secondary linkages are not strong bonds, like those holding the polypeptide chains together. They are weak — but they are many. Now when we talk about bonds being weak in a structure, we mean that they are readily disrupted by mild conditions. Thus, proteins are often markedly delicate structures — like life itself — and unable to withstand harsh treatments. The white of an egg, for example, consists of the protein albumin; and simple heating, as we all know from everyday life, irrevocably changes its properties. The identity of the protein as a molecular entity is thus conferred on it by the strong peptide links holding together the amino acids. But its properties — the qualities that make it essential for life — arise from the multiple weak interactions that create the native structure.

The analogy is often made that the 20 amino acids may be considered like the letters of an alphabet. A protein is formed from the particular sequence of letters. This analogy applies only to a word, as it were, in an indecipherable language. When the polypeptide chains of a protein are in a simple linear or one-dimensional arrangement, the protein is called "denatured" or "unfolded" and it does not have its specific properties. The sequence does not display its "meaning." Only when folded in space into their native conformation do these polymers display their essential characteristics. Thus, not merely the *informational* content of the word — that is, the sequence of letters — carries its meaning, but also its three-dimensional structure. An arrangement of letters without a translation is therefore something like an unfolded polypeptide chain of a protein.

A protein molecule may be made up of more than one polypeptide chain — usually there are just a few — and each chain has a specific sequence of amino acids. Biochemists have worked out methods for determining the number of chains and the ordering of amino acids within a chain. Now, the magnitude of the problem of protein structure may be illustrated by the fact that the first protein sequence to be determined was that of insulin. This is the hormone whose deficiency gives rise to the disease diabetes. Insulin is a very small protein with 51 amino acids in two chains, which are cross-linked to one another (figure 2). The sequence was first determined by Sanger in 1953. This work was a milestone in our understanding of proteins since it showed, unequivocally, that each kind of protein could be characterized by a specific sequence of amino acids. Thus, it proved that proteins really were molecules! Nevertheless, the three-dimensional structure of insulin — that is, the specific folding of the molecule — could not be predicted from this sequence, and therefore the mechanism of action of the hormone could not be approached from a structural point of view. Only in 1971, twenty

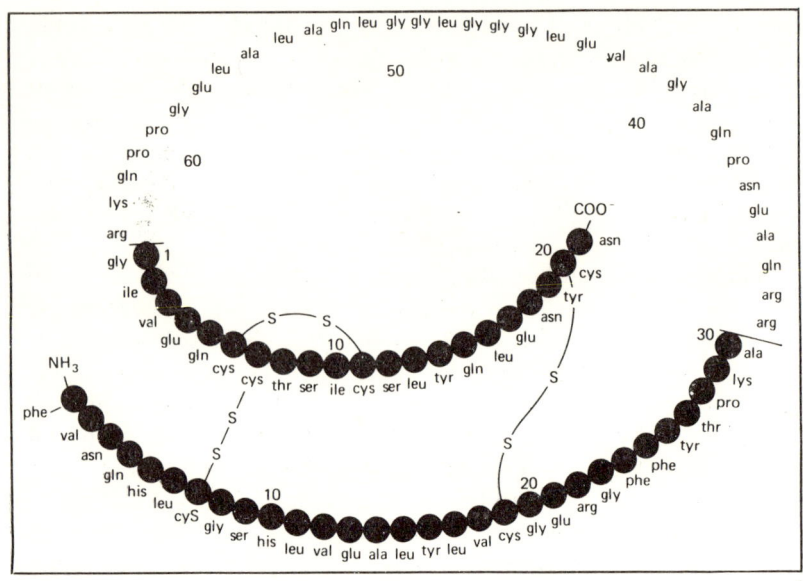

**Figure 2.** Insulin sequence. The primary structure of porcine proinsulin (removal of segment 31-63 converts the molecule to insulin). Wold, Finn (1971). *Macromolecules: Structure and Function.* Prentice-Hall, New Jersey. From *Science 161,* 165. Reprinted with permission of *Science* and Dr. R.E. Chance. Copyright © 1968 by the American Association for the Advancement of Science.

years after the sequence was determined, has the folding of insulin been established.

The solution of the structure of proteins has been one of the great riddles of modern biology. The answer certainly lay beyond the knowledge of amino acid composition — and even sequence — that is, beyond the domain of organic chemistry. The answer required, in fact, a means of seeing atoms in space — and here new developments in physics opened the way. J.D. Bernal, whom many of us consider to be the founder of modern molecular biology, was one of the physicists who focussed attention on the importance of determining the three-dimensional structure of proteins. So long as the structure of proteins was a mystery, the nature of life itself could not be understood. Bernal was a close colleague of Dorothy Wrinch, and she shared his belief in the deep significance of this problem. Now, Bernal was a crystallographer; in fact, a pioneer in the application of x-ray crystallography to biological structures. And it was the enthusiasm of British crystallographers for this most difficult problem that led to many of the advances that culminated in the actual discoveries of how proteins are built.

The solving of the riddle of protein structure by x-ray diffraction is a drama extending over many decades. The solution of the three-dimensional structure of insulin is but one of the many recent triumphs in protein crystallography. At present, about 70 protein structures have been solved to atomic resolution. The first of these were the fibrous proteins.

## X-RAYS REVEAL STRUCTURE

The solution to the structure of proteins required the use of x-rays.[7] It required, in fact, an understanding of how the scattering of x-rays from a substance can be interpreted to show the arrangement of its atoms. X-rays had been discovered by Roentgen in 1895, but for many years their nature and usefulness were not understood. Nor in those days was the nature of crystals understood. Roentgen was the Director of the Institute of Experimental Physics at the University of Munich, which, at that time, had an extraordinary constellation of talent. Thus, Arnold Sommerfeld, an expert on optics, was the director of the group in theoretical physics. One of Sommerfeld's students, Ewald, had chosen a Ph.D. thesis topic in crystal optics. Ewald discussed this problem with Von Laue, who had joined Sommerfeld and who was then writing an article for an encyclopedia on the optical properties of gratings. Following these talks, Von Laue had the inspiration that crystals might act as gratings for x-rays: just as ruled gratings scatter visible light in particular directions, so ordered arrangements of atoms in crystals might scatter x-rays (figure 3). This would mean that x-rays were radiation of very short wavelength — about the size of the distance between atoms. So the question was asked: could crystals "diffract" x-rays? And now the classic experiment was performed. It was performed, I might add, by a few stubborn experimentalists against the advice of a number of theoretical physicists, all of whom had produced very good reasons why it could not work. But it did work. A narrow beam of x-rays was aimed at a crystal, and a photographic plate detected the scattering from the crystal (figure 4). Crystals *do* diffract x-rays. This result produced a whole new field — and a number of Nobel prizes. It revealed the nature of x-rays as radiation of short wavelengths, and it provided a tool for investigating the internal structure of crystals.

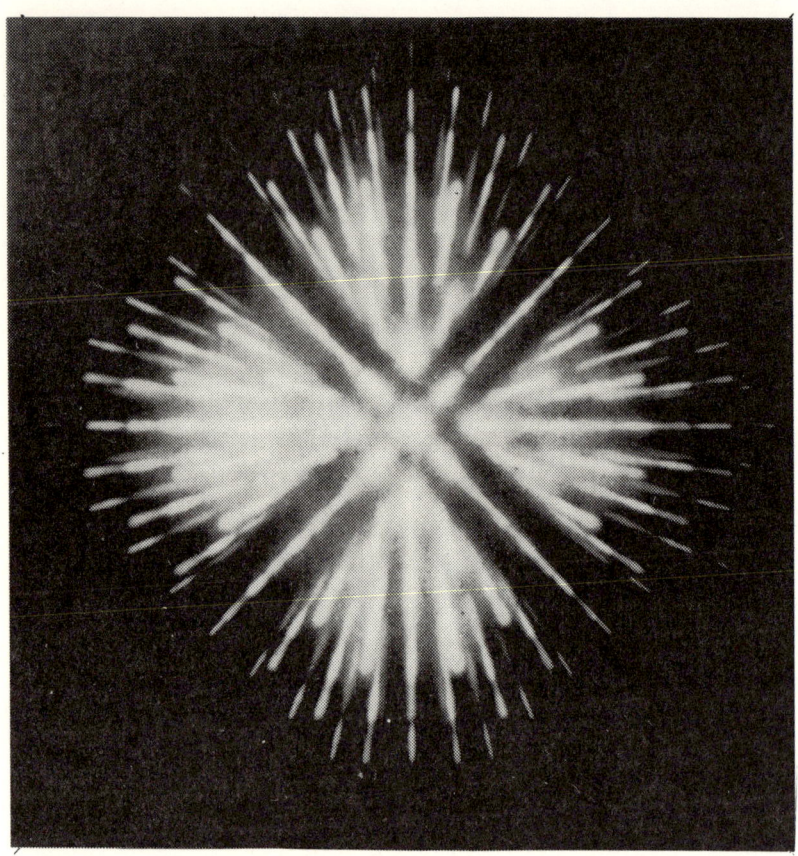

**Figure 3.** Diffraction from grating with white light. A similar effect is seen by viewing a distant light through a piece of fine material. The pattern of fringes would show colors not illustrated here. Ronchi, Vasco (1970), *The Nature of Light* (trans. V. Barocas), Harvard University Press, Cambridge, Mass. Reprinted with permission from Harvard University Press.

When news of the discovery reached England, W.H. Bragg, in Leeds, and his son Lawrence, who was a student at Cambridge, set to work on its implications. They deduced the composition of x-ray beams, the scattering properties of atoms, and Lawrence Bragg conceived of a simple relation between diffraction effects and the geometry of crystals. This is the well-known Bragg Law. X-rays have very short wavelengths, and there are no good lenses to focus them, so that we don't have useful x-ray microscopes. The same property,

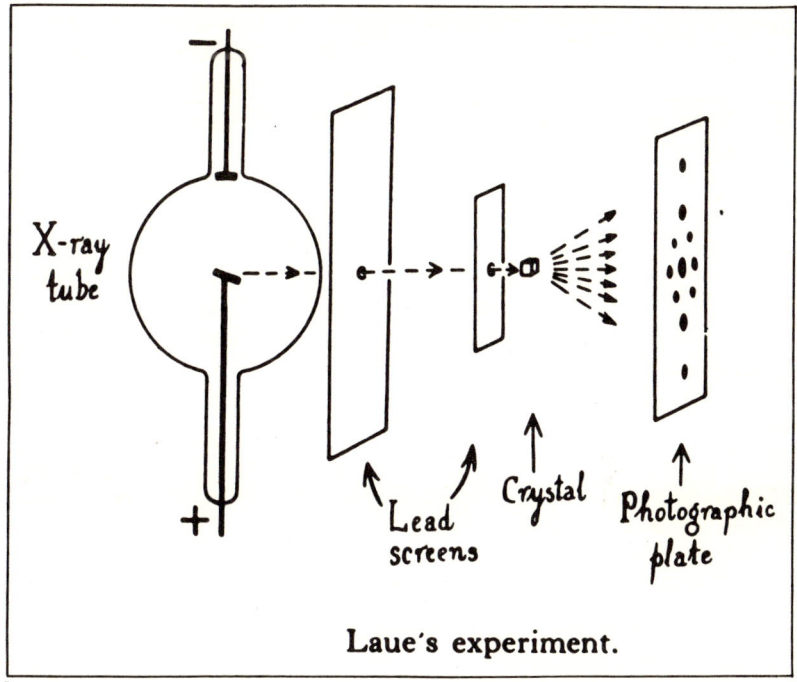

**Figure 4.** First x-ray diffraction experiment: Laue's discovery. Diagram of arrangement used by Friedrich and Knipping. A beam of x-rays (limited by lead screens) strikes a crystal and the scattered rays are recorded on a photographic plate.

however, means that x-rays are capable of revealing the arrangement of atoms — but only if one is able to interpret their scattering. Both father and son subsequently developed the methods for using x-rays to determine the structure of matter. They were awarded the Nobel Prize in 1915 — at the time the son was at the front. The Braggs essentially founded the science of x-ray crystallography. Lawrence Bragg had the great pleasure of seeing successive triumphs of x-ray crystallography: from the solution of the structure of simple salts, to minerals, to organic compounds, and, finally to the most awesome problem of all — the arrangement in space of the atoms in the giant molecules of proteins.

## FIBROUS PROTEINS ARE HELICES

W.H. Bragg became Director of the Royal Institution in London in 1923 and began the analysis by x-rays of simple organic molecules. At that time, W.T. Astbury — as well as J.D. Bernal — was one of the early workers in the laboratory. Bragg had marvelous powers of exposition, and his Friday Evening Discourses — which were popular presentations about science — were justly famous. For one of these lectures, titled "The Imperfect Crystallization of Common Things," he asked Astbury to take x-ray photographs of various biological *fibers*, such as hair, horn and muscle. The results so interested Astbury that he made this field his life's work, later going to Leeds to start the first school of Biomolecular Structure. It was Astbury, in fact, who first used the term molecular biology, and he had a very clear conception of its meaning.

In examining protein fibers, Astbury discovered that these systems gave essentially three kinds of x-ray patterns.[8] Such diverse proteins as animal hair, the epidermis of skin, muscle, and the clotting protein from blood plasma gave a characteristic pattern, or

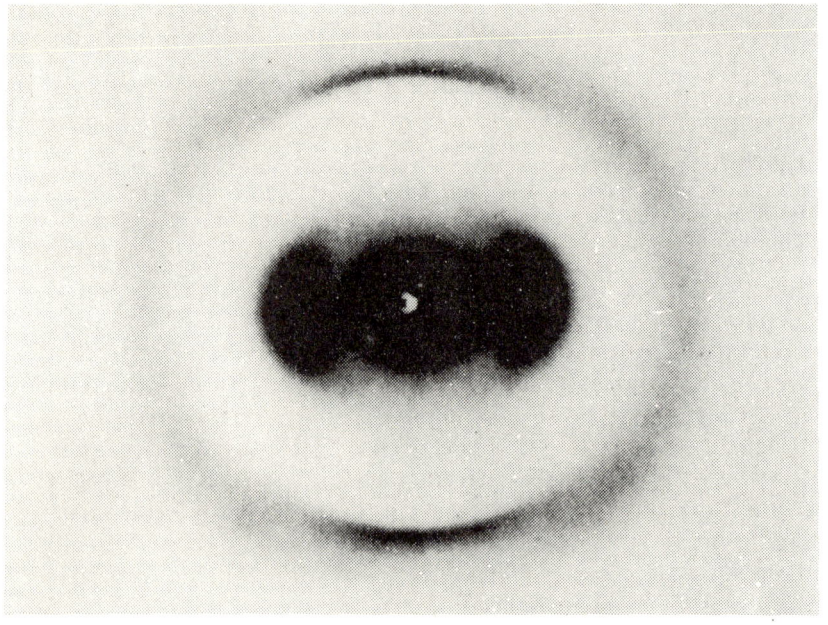

**Figure 5.** $\alpha$-pattern. This x-ray diagram is from a fiber of muscle proteins. The discrete spots show that there is an orderly arrangement of the chains in the molecule. *Photograph by C. Cohen.*

x-ray "fingerprint" (figure 5). Astbury called the polypeptide chain structure in this class of proteins the α fold.

He also found that hard reptilian scales, feathers and silks gave another pattern. He called the chain conformation which gave this diagram the β fold (figure 6). A similar pattern could also be ob-

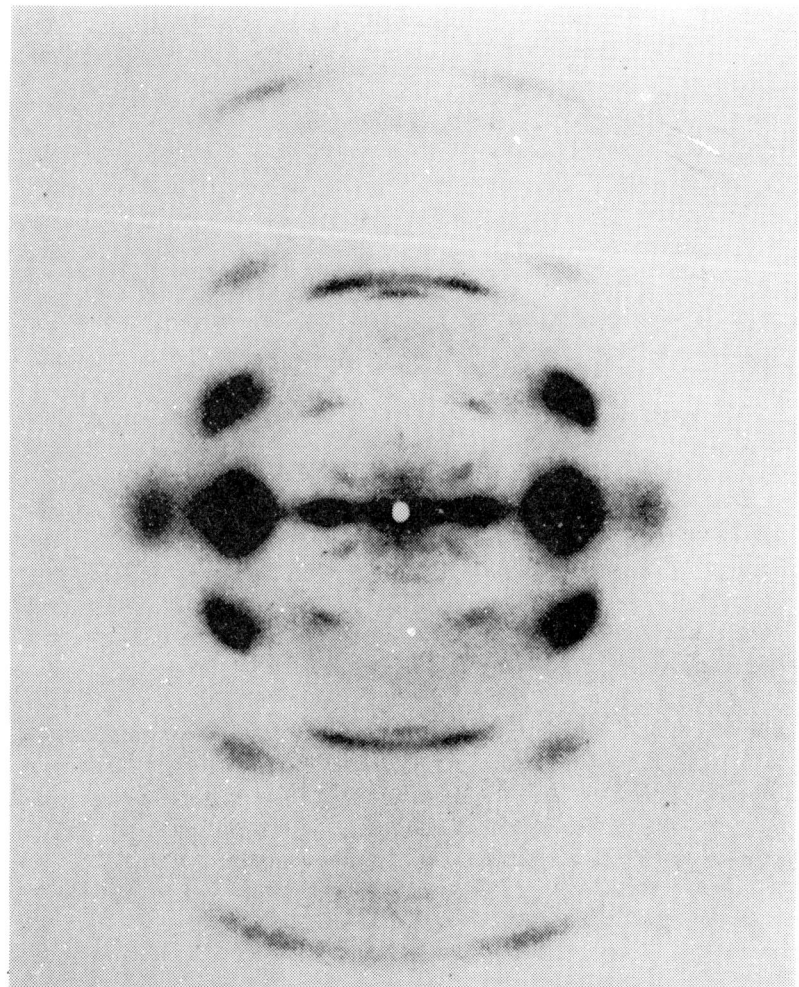

**Figure 6.** β-pattern. The x-ray pattern of silks shows that the chains are ordered, and almost fully extended. Marsh, R.E., Corey, R.B. and Pauling, L. (1955), *Biochim. Biophys. Acta 16*, 1. Reprinted with permission of Elsevier/North-Holland Biomedical Press.

tained merely by stretching the α proteins. Thus, hair and wool are elastic, as one knows from attempting to preserve the shape, for example of a wool sweater. It's the α polypeptide fold that one is protecting! When extended, these proteins show the β fold.

The third kind of pattern was that given by the protein collagen, which makes up connective tissue, tendon and bone. In fact, collagen makes up the largest proportion of protein in the human body. Gelatin is derived from collagen by boiling the protein and so unfolding or "denaturing" its polypeptide chains.

All these fibrous proteins yield meager or poor x-ray diagrams. But the simplicity of the diagram shows an essential simplicity of the folding of the polypeptide chains. Astbury did not solve the structure of any of these proteins; but he set the problem, as it were, and clearly signalled the implications of the solution.

The structures of the fibrous proteins were solved by building models — because these structures are essentially regular — and by recognizing how helices diffract x-rays. It has turned out that fibrous proteins are helices. A helix is a spiral like a spring. And it is the most *natural* kind of structure for fibers. In a helix, each unit makes the same linkages as each of its neighbors. Helical structures had been suggested for polypeptides in the 1940s by Huggins, for example, but the structural chemistry of polypeptides had not advanced sufficiently to lend physical plausibility to the models. Moreover, there was no simple way to test the models by diagnostic x-ray diffraction patterns. In the beginning of the 1950s, these obstacles were dramatically overcome.

In 1950, Bragg, Kendrew and Perutz published a paper describing a systematic theoretical search for possible helical polypeptide structures.[9] Their approach was too restrictive in one key aspect and too relaxed in another. They adhered to formal crystallographic concepts in allowing the helices to have only an integral number of amino acid residues in each turn; that is, no fractional parts of a residue were allowed per turn. On the other hand, they didn't have sufficient experimental evidence to know just how flexible the chain might be. Thus, a number of structures couldn't be ruled out — yet they missed the discovery of the α-*helix*.

Pauling announced the discovery of the α-helix in 1951;[10] he had recognized one aspect of the complex "inner logic" of protein chains (figure 7). Following major findings on the nature of the chemical

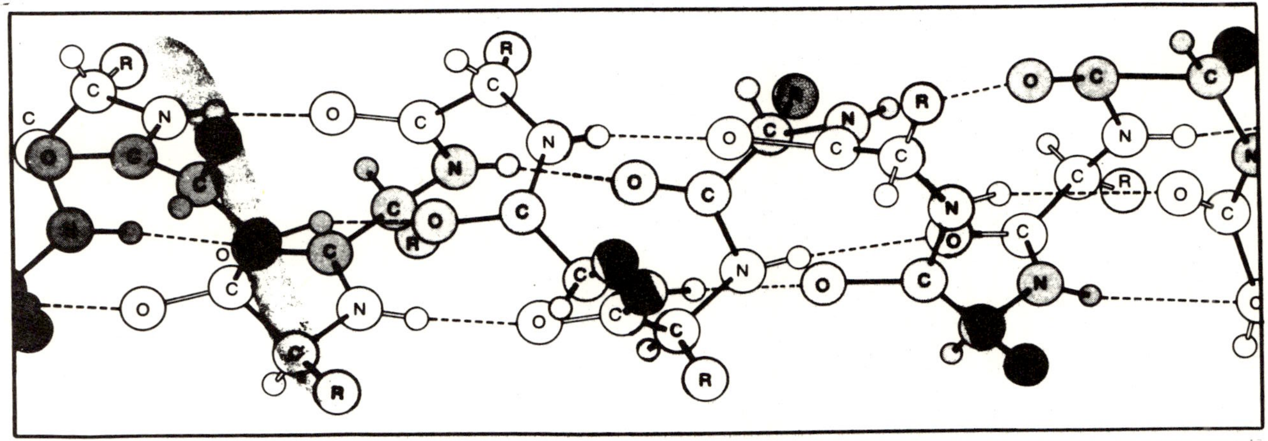

Deciphering Protein Designs 89

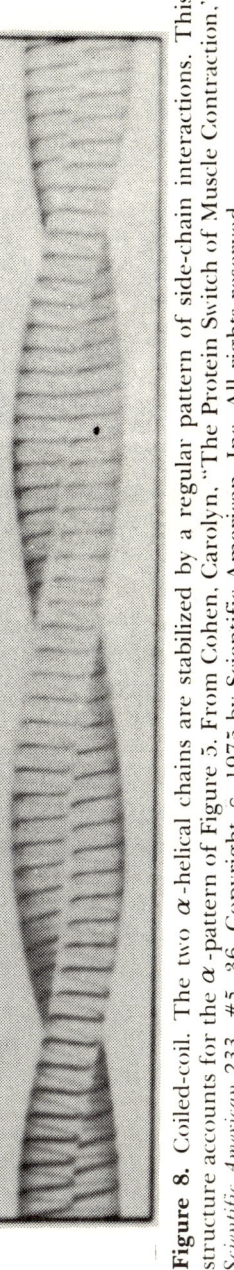

**Figure 7.** α-helix. Successive turns are held together by hydrogen bonds. The helix is right-handed in nature. From Cohen, Carolyn, "The Protein Switch of Muscle Contraction," *Scientific American* 233, #5, 36. Copyright © 1975 by Scientific American, Inc. All rights reserved.

**Figure 8.** Coiled-coil. The two α-helical chains are stabilized by a regular pattern of side-chain interactions. This structure accounts for the α-pattern of Figure 5. From Cohen, Carolyn, "The Protein Switch of Muscle Contraction," *Scientific American* 233, #5, 36. Copyright © 1975 by Scientific American, Inc. All rights reserved.

bond, where he used results from the x-ray analysis of minerals, Pauling began a decade of work on the crystallography of amino acids and peptides with R. Corey. From these studies, Pauling saw the crucial constraint that the peptide unit be planar. It meant that "there was only one place where a corner can be turned in folding the chain,"[11] This is probably the key conceptual innovation. Moreover, from the earlier work he recognized that the helix would be most stable if each turn of the helix was linked to the next by the maximum number of what are known as "hydrogen bonds." Hydrogen bonds are a special kind of very weak link, which gives water, for example, its unusual properties. Pauling describes his discovery of the $\alpha$-helix as taking place in Oxford, England when he was recovering from a cold and confined to bed. He was so bored with mystery stories that he began building models of protein helices. And using these restrictions, he drew out the $\alpha$-helix on a folded sheet of paper. The structure, in effect, built itself. For the strongest hydrogen bonding it had not 3 or 4 amino acids, but 3.6 residues per turn. It was elegant, it was chemically sound, it was bound to be right.

The $\alpha$-helix is a beautifully simple structure, and it quickly became possible to test its reality. Crystallographers soon worked out the "language" of helical diffraction patterns.[12] The simplicity of this kind of structure meant that the x-ray diagrams they gave could readily be recognized. Chemists then synthesized polypeptides that were as similar to proteins as possible. That is, there were many amino acids in the chain, but generally of only a few types, and the $\alpha$-helix was shown to account exactly for their x-ray patterns.

Now, the $\alpha$-helix was right for synthetic polypeptides, but proteins were not quite so simple. Perutz, in England, showed that there was a close connection between the x-ray diagrams from synthetic polypeptides and those from Astbury's $\alpha$ proteins, but that they were not identical. The solution to this discrepancy was given by both Pauling and Corey and by Francis Crick. Crick pointed out that the geometry of the $\alpha$-helix was such that two $\alpha$-helices cannot pack together properly with optimal meshing of side chains unless they coil round one another. In so doing, they form a larger helix, or *super-coil*, and there can be systematic weak linkages between the side chains of two helices[13] (figure 8). In water solutions, where many of these proteins are found, the nonpolar or oily side chains would "zip up" the two chains at the center and the charged groups would be at the surface interacting with water and conferring solubility to the proteins. This suggestion of super-coils predicted new features of the x-ray pattern, which were found.[14] The solution to one class of proteins had finally been established.

The collagen fold is an analogue of the $\alpha$-helical super-coil. It was solved by knowing how helices diffract, by model building and by work on synthetic polypeptides. The reason model building was successful in this case is that the amino acid sequence of collagen is most unusual. Glycine, which has as its side chain only the very small hydrogen atom, occurs precisely every three residues. Moreover, the protein collagen is characterized by an extraordinarily high proportion of the amino acids proline and hydroxyproline whose side chains simply do not fit neatly into an $\alpha$-helix. Collagen turned out to be a triple-stranded super-coil. The three chains "co-crystallize" around a common axis, and a set of regular interactions between chains is established. Each of the individual chains of the collagen triple-stranded structure is itself a helix, and the three chains wind round one another with the special residue glycine located at the center, where only it can fit.

Now, you will remember that the $\beta$ structures occur naturally, and are also formed by simply stretching $\alpha$ proteins. So one knows that in the $\beta$ form the polypeptide chains are pretty well extended. Pauling and Corey suggested a sheetlike structure for the $\beta$ conformation[15] (figure 9). The polypeptide chains are essentially uncoiled, and aligned to make two sets of weak linkages: in one direction the backbone atoms bond between chains to form a sheet structure; between sheets there is a regular meshing of side chains. The strength and quality of silk are well accounted for by this structure: silks are relatively inextensible since the chains are almost fully stretched; weaker linkages between sheets make the fibers flexible. Feather can be shown to be based on a similar plan, but it is a somewhat more complicated structure.

The fibrous proteins thus illustrate three basic types of polypeptide chain folds: the $\alpha$, the $\beta$ and the collagen-fold. Recent calculations indicate that these are conformations that are energetically favorable; that is, they can now be predicted to be stable conformations. The x-ray patterns, as I mentioned before, were not very detailed and showed a good deal of disorder in these proteins. This means that the structures cannot be solved in detail, *i.e.* to show their atomic parameters. But they do illustrate the following principles of protein structure: that essentially three favored polypeptide chain conformations exist depending on amino acid composition

92   *Structures of Matter and Patterns in Science*

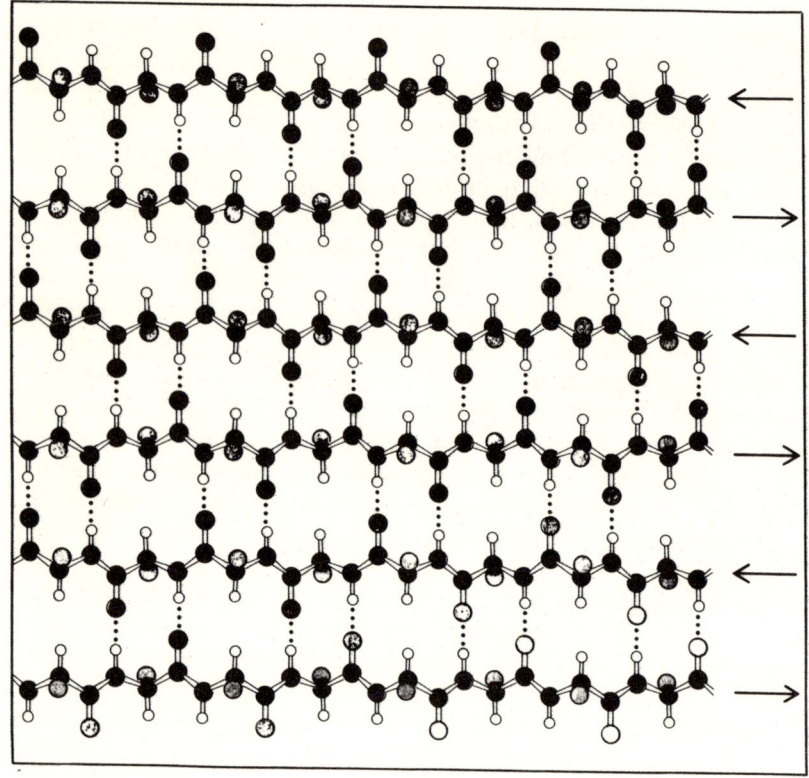

**Figure 9.** $\beta$-pleated sheet. In this model adjacent strands run in opposite directions. There is hydrogen bonding in the plane of the sheet. This kind of structure accounts for the $\beta$-pattern (Figure 6) given by such fibers as silks. Stryer, L. (1975); op. cit. Reprinted with permission from *Biochemistry* by Lubert Stryer, Copyright © 1975, by W.H. Freeman and Company.

and sequence. None of these proteins is a single chain structure: weak linkages between chains stabilize these folds, with specific side chain interactions playing a key role; that the nonpolar side chains will tend to pack together away from contact with water, while charged groups will tend to be at the surface. These principles have been demonstrated in greatest detail in the recent discoveries on the structure of globular proteins.

# THE RIDDLE OF GLOBULAR PROTEINS

Globular proteins are often called "crystalline proteins" — and they do, indeed, form beautiful crystals (figure 10). These had been

**Figure 10.** Myoglobin crystals. These are reddish in color, due to the heme groups in the molecule. Myoglobin is present in muscle and gives meat its color. Reprinted with permission of Dr. J.C. Kendrew, from *The Thread of Life*, Harvard University Press, Cambridge, Mass., 1968.

seen as early as the 1800s, but their nature was not understood. The discovery that proteins form crystals is usually credited to J.B. Sumner, who crystallized the enzyme urease in 1926 and showed that it was a protein. At the time, both findings were important breakthroughs. A number of globular proteins, many of them enzymes, were then quickly crystallized, and in 1934 a sample of the crystalline protein pepsin was put into the hands of J.D. Bernal. Now, Bernal had gone to Cambridge University from the Royal Institution, but he had taken with him the tradition of the latter place, where, in his words, "crazy experiments were the rule."[16] When he attempted to take an x-ray diffraction photograph of one of these crystals, he observed only diffuse scattering on the film. But the crystal had dried in the course of the exposure. Bernal then had the idea of sealing a bit of the solution from which the protein had been crystallized in a thin-walled glass tube with the crystal. When mounted in this wet state, the crystals gave beautifully detailed diffraction patterns (figure 11). To quote from his student at the

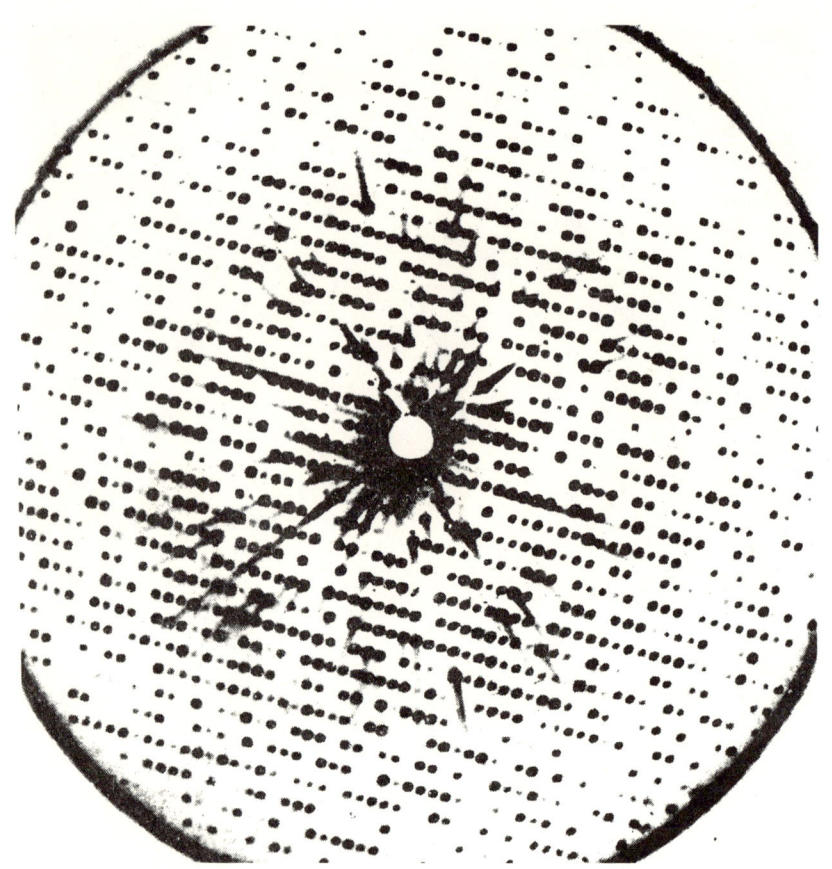

**Figure 11.** X-ray pattern from myoglobin crystal. This rich pattern of spots shows the high degree of order present in the myoglobin molecules. But the phase information necessary to solve the structure is lacking in this pattern. Reprinted with permission of Dr. J.C. Kendrew, from *The Thread of Life*, Harvard University Press, Cambridge, Mass., 1968.

time, Dorothy Crowfoot-Hodgkin, "... that night, Bernal, full of excitement, wandered around the streets of Cambridge, thinking of the future and of how much it might be possible to know about the structure of proteins if the photographs he had just taken could be interpreted in every detail."[17] Shortly thereafter, a number of other protein crystals were shown to give excellent x-ray patterns. Yet, it took more than 20 years for the way to solve these patterns to be discovered.

In contrast to their arrangement in fibrous proteins, the polypeptide chains in globular proteins are not simply coiled. In scientific jargon, the chains are not "regularly periodic." Yet they are folded into very well defined structures and that is why, in fact, they crystallize. Moreover, that is why the x-ray diagrams of these crystals are so detailed. Now, there is a paradox here. A poorer x-ray diagram can, in some cases, give more information than a detailed one. Thus, the fibrous proteins do not generally crystallize and the x-ray patterns given by these proteins are rather diffuse. Yet, the diagrams show that the molecules are helical and give the basic rules for building models for the structures. In contrast, the globular proteins form highly ordered arrays in three dimensions; that is, crystals. And these crystals give x-ray patterns showing very many discrete spots or reflections. But, because of the very order of the crystal, the scattering from the individual molecules is not easily seen. It is a bit like not seeing the trees for the forest. And even if the diffraction, or scattering, from the individual globular protein molecule could be seen, there is no way that the folding could be recognized from the complex scattering pattern since, unlike simple helices, the chains are not wound in any regular way. Thus, the globular proteins yield x-ray patterns with vastly more information encoded in them than do the fibrous proteins, but it was not easy task to break this code.

In order to solve a pattern of this kind, an image has to be calculated from the intensity and position of the spots on the x-ray diagram of the crystal. The crystaliographer produces this image by a mathematical procedure in precise analogy to what the lens does physically with comparable information from light rays. But there is one crucial piece of information missing from the x-ray diagrams that must be added before an image can be formed. The different atoms in a structure scatter x-rays, which reinforce or cancel one another; that is, the x-rays interfere with one another depending upon the precise position of the atoms. This information is called the "phase" of the spot on the x-ray pattern. Now, the film can only record intensity — that is, energy — and the relative time of arrival of the x-rays, or "phase" information, is thus lost. It is not directly available from the pattern and must be deduced or determined in some other way. This missing information, which is available to the lens in forming an image by light rays, constitutes the major problem in solving the x-ray diagrams from the crystalline proteins. Now, for relatively simple substances containing a small number of atoms, the phases can be obtained by model-building or by trial and error, or by other techniques. And such methods could be extended rather far so that the largest structure solved by this approach was vitamin B12, which has 181 atoms. It was solved by Dorothy Crowfoot-Hodgkin and collaborators only in 1956, after eight years of work. And it was a triumph of classical crystallography. But a small protein has thousands of atoms. Hence, the crystals yield thousands of relatively weak x-ray reflections, whose intensity must be measured and added to one another with their correct *phase* to produce the structure. That truly is a Herculean task.

Long before the phase problem was solved, and, indeed, since the earliest recognition that globular proteins are molecules with well-defined structures, there were attempts to define the structure by what we may call "inspired guesses." This was, in a sense, an approach by model-building. But the underlying assumption in all these attempts was that the structure of the globular proteins showed some kind of regularity. Without this assumption, there could be no key to the solution. And then various experimental data were examined to see whether these inspired guesses could be verified. Scientists were, in fact, looking for a shortcut to the solution of the globular proteins. Rembrandt's famous etching — the so-called "Faust" — symbolizes this aspiration (figure 12). To quote Rosenberg,[18] "We see a scholar who has risen from his desk at the appearance of an extraordinary vision in his window. He had been deep in his work . . . The eye travels to the brilliantly lighted head of the scholar and on to the disk, which finally holds our attention as the focal point in the pictorial organization . . . In the disk it seems lies the real source of the unusual illumination in spite of the large window above." This beautiful and mysterious print, showing a scholar experiencing a supernatural vision, can be viewed as the scientist solving — in one inspired guess — a riddle about which he had long puzzled.

There are a number of examples of early attempts to reveal a basic plan in globular proteins. One of the most elegant hypotheses was proposed by Dorothy Wrinch. She pictured amino acid residues linked in such a way that they formed hexagonal rings ("cyclols"). These patterns could be extended and folded. Thus, she envisaged globular proteins as comprising a series of space enclosing molecules[19] (figure 13). The cyclol theory gained early support and enthusiasm for its beauty and brilliance, but it was then criticized on

**Figure 12.** Rembrandt "Faust" etching. Originally entitled "Practisierende Alchimist." From *Rembrandt: Life and Work* by J. Rosenberg (Plate 234). Reprinted with permission of Phaidon Press, Ltd., Oxford. (Bartsch 270).

chemical grounds; the structure did not have the peptide linkage, which seemed clearly to be found in proteins. Nevertheless, as John Edsall has said, "... it was not until Wrinch developed a very de-

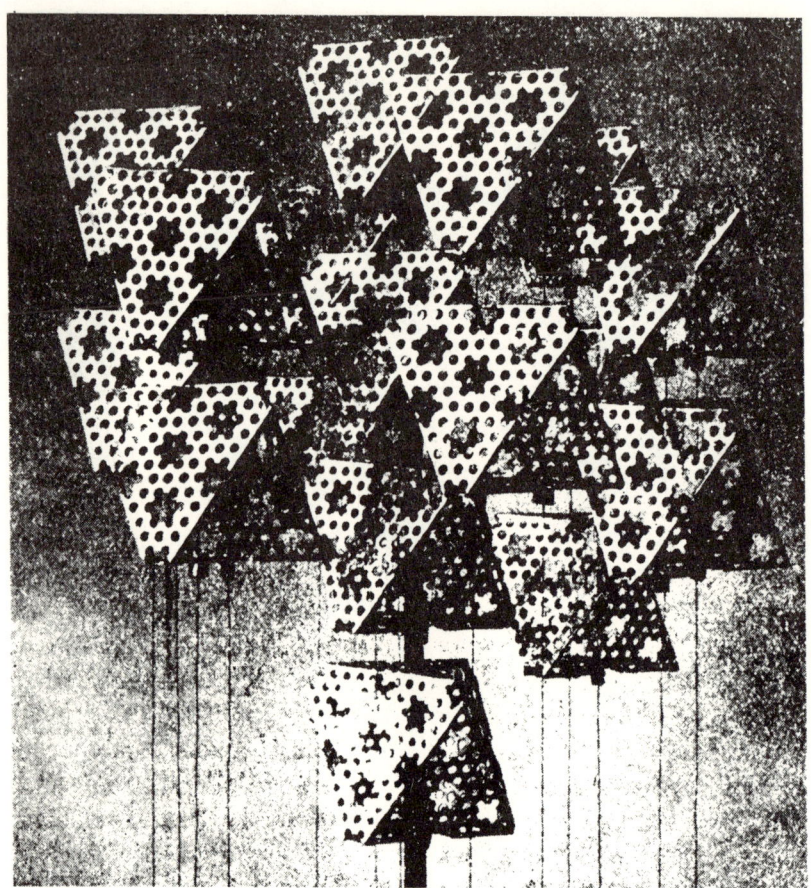

**Figure 13.** Wrinch cyclol model of a zinc-insulin crystal. Wrinch, D. (1941). *Phil. Mag.*, Ser. 7, XXXI, 177. Reprinted with permission of Taylor and Francis, Ltd., London.

tailed and explicitly geometrical argument that structures of this sort were seriously debated by chemists."* Moreover, this idea also stimulated Wrinch to extend the theoretical analysis of x-ray diagrams of globular proteins. These crystal patterns were like the untranslated Rosetta Stone, and her efforts were directed toward deciphering the difficult "language" of the diffraction diagram.[20]

An additional example of the model-building approach to the solution of the globular proteins is that involving the pioneering work on the structure of hemoglobin and myoglobin, undertaken at

* See note 5.

Cambridge University. In 1935, Max Perutz came from Vienna and joined the laboratory of J.D. Bernal. There he began his x-ray crystallographic studies of hemoglobin. This is the iron-containing protein which carries oxygen in blood. At the time Perutz chose the structure of hemoglobin for his thesis, the largest structure that had been solved by x-rays contained only 58 atoms. Hemoglobin was more than 100 times larger than this. After the war, and due in part to the encouragement of Bernal, John Kendrew became a graduate student of Perutz's at Cambridge and chose for his thesis the x-ray structure of myoglobin — the so-called "baby brother" of hemoglobin. This is the oxygen-carrying protein in muscle. Now, myoglobin itself was a formidable problem since it contains 2500 atoms. And so began at the Cavendish Laboratory, the long "25-year loneliness of the protein crystallographers."[21] Before success was to be theirs, many years of often frustrating work were to be spent attempting to decipher the x-ray diagrams of hemoglobin and myoglobin. At the time, there was no way of going objectively from the patterns directly to the structure. Neither Perutz nor Kendrew seriously believed that the structures could be solved by model building, but in attempting to interpret certain features of the intensities on the x-ray diagrams, they had to make what were then plausible assumptions. Thus, they tried to discern the presence of short lengths of rods — or polypeptide chains — which, when grouped together, would form the structure of the globular protein. Similarly, other protein crystallographers also searched for parallel rods in such proteins as insulin and ribonuclease. In imposing this notion of regularity on the globular proteins, the crystallographers were badly misled. The correct solution came only by using a new method to interpret the x-ray patterns. And the solution revealed that the protein molecules were "something quite new in science."[22]

The long loneliness finally came to an end. Perutz and his colleagues discovered that a heavy atom — that is, the atom of a metal like mercury — could be attached to the protein molecules in a crystal without disturbing the packing of these molecules. This single atom then produced changes in the intensities on the x-ray pattern. Thus, the fact that proteins were weakly diffracting was an advantage since the presence of a single heavy atom specifically labelling the protein could be discerned. These intensity changes enable one to locate the position of the heavy atom in the molecule and this fact in turn allows one to overcome the phase problem.[23]

Attachment of a number of such heavy atoms, such as uranium and lead, will, in general, be necessary for the complete solution of the protein structure. Moreover, this key discovery was made at a time when high-speed computers had been developed. Computers were essential for the calculations since there were so many data to be processed.

The first protein structure to be solved was myoglobin in 1957.[24] The pattern has, in all, about 50,000 reflections so the analysis had to

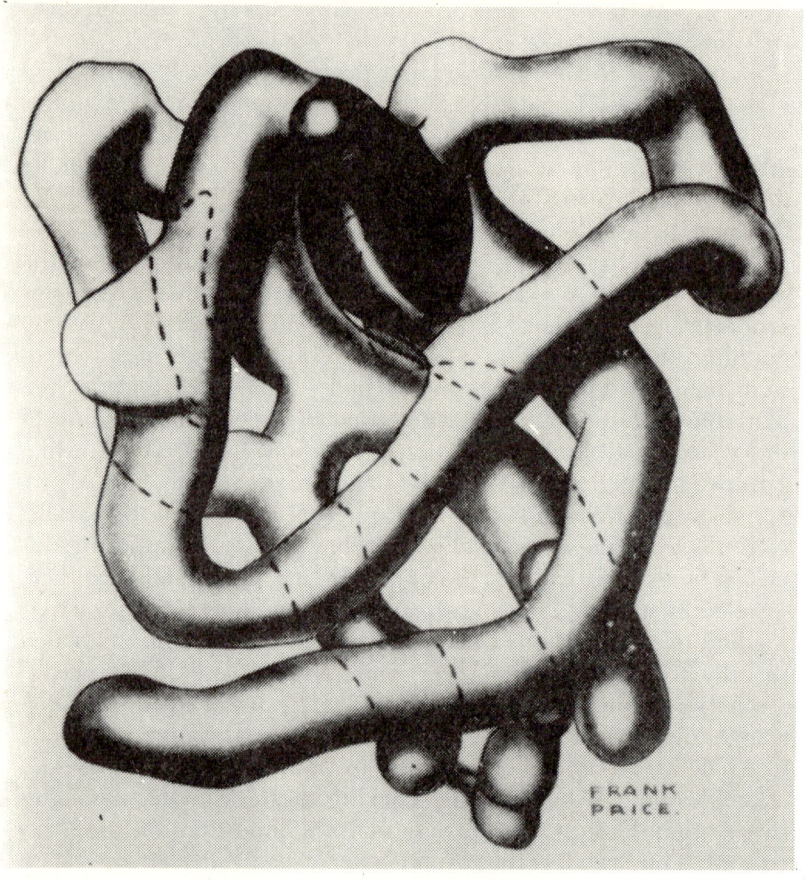

**Figure 14.** Myoglobin structure at low resolution. The polypeptide chain is seen as an irregularly bent rod forming a container for the heme group (black disk). Reprinted with permission of Dr. J.C. Kendrew, from *The Thread of Life*, Harvard University Press, Cambridge, Mass., 1968.

proceed by stages. The first image used only a small fraction of the available data and a so-called "low-resolution map" was obtained. This is a picture of the molecule with just enough detail so that it is possible to trace the polypeptide chain throughout the protein and to show the location of the iron atom (figure 14). The most striking fact about the structure of myoglobin was that the polypeptide chains were not arranged in a regular or symmetrical pattern, but that they were incredibly folded and contorted. It was, in the classical sense, not a beautiful molecule at all. The cyclols of Wrinch were symmetrical creations of the human mind. The structures created by nature were far more complex.[25] It was inconceivable that the actual structures could have been deduced by any inspired guess.

In order to see the precise coiling of the polypeptide chains in the structure, a higher resolution picture had to be obtained, which

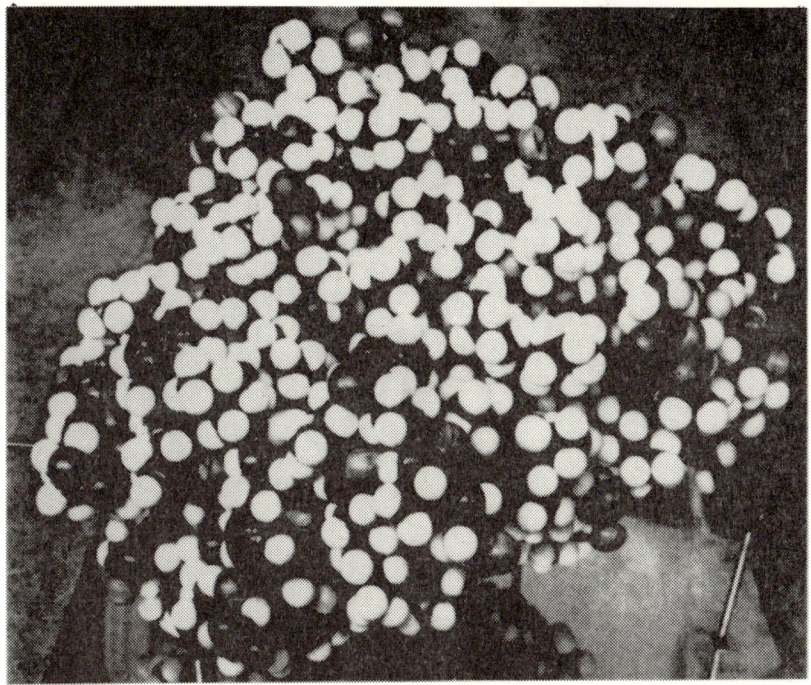

**Figure 15.** Myoglobin structure at atomic resolution. This space-filling model (built by Dr. H.C. Watson) shows all 2500 atoms intricately packed to form the molecule. Reprinted with permission of Dr. H.C. Watson from Kendrew, *The Thread of Life,* Harvard University Press, Cambridge, Mass., 1968.

meant measuring many more spots and adding them together to produce an image. Thus, in 1959, individual atoms could be seen and by 1962 a high resolution structure was obtained (figure 15).

These more detailed images revealed $\alpha$-helices! They showed for the first time that $\alpha$-helices conceived of by Pauling did really exist in globular proteins, as well as in fibrous proteins. They showed, moreover, that the structure had a number of corners, which were turned in quite different ways. The interior of the molecules was tightly packed with nonpolar residues and most of the amino acids with charged groups were at the surface. This kind of structure has

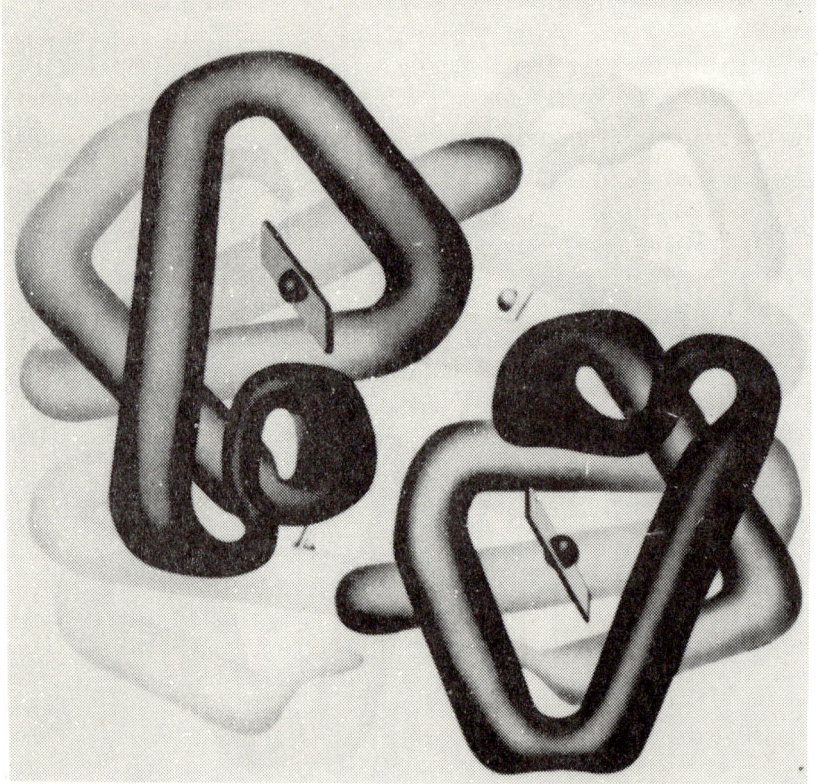

**Figure 16.** Hemoglobin structure at low resolution. The two pairs of $\alpha$ and $\beta$ chains are themselves folded similarly, and very like myoglobin. The four subunits make up one molecule. Reprinted with permission from *The Structure and Action of Proteins* by Richard E. Dickerson and Irving Geis. W.A. Benjamin, Inc., Menlo Park, California, Publisher. Copyright © 1969 by Dickerson and Geis.

been described as a sort of "oil drop" with the hydrocarbon chains at the center and polar groups outside. This is perhaps, the clearest generalization that has emerged so far about the relation between amino acid sequence and three-dimensional structure. Another striking fact derived from comparative studies of myoglobins of different species was that there may be a wide variety of substitutions at any site in the interior of the protein; for example, there is no simple rule determining which nonpolar residues appear at a given position. But more about this later.

Hemoglobin is four times as large as myoglobin and more complicated. And it was not until the summer of 1959, "nearly 22 years after Perutz had taken the first pictures of hemoglobin,"[26] that its structure could be seen. The extraordinary finding was that each of the four chains of hemoglobin looked very much like a myoglobin molecule (figure 16). Hemoglobin, in fact, consisted of one pair of chains almost identical with myoglobin, and another pair of chains differing by only one small loop. Now, an interesting fact about myoglobin and hemoglobin, the first two globular proteins whose structure was determined, is that they contain a very high proportion — roughly 70-80% — of $\alpha$-helix, and this very large amount of helix was helpful in tracing the chain from the early low resolution results.

The next protein whose structure was determined was the enzyme lysozyme, solved in 1965 by D.C. Phillips and his collaborators (figure 17.)[27] At that time, additional phase-determining methods were in use. Other structures, including those of a number of enzymes, such as ribonuclease, chymotrypsin and so forth have followed rapidly since 1967. Enzymes are the specific catalysts of cells promoting reactions which otherwise could not take place. In addition to the $\alpha$-helix, the $\beta$-pleated sheet is a chain conformation which has been found in many globular proteins. The $\beta$-sheet is straight in a fibrous protein such as silk, but in a large variety of globular proteins the sheets have an additional twist. One regular structure commonly found is called the $\beta$-barrel.[28] Here a cylinder is formed, made up of staves or polypeptide strands of $\beta$-pleated sheet, which are inclined to the axis of the barrel (figure 18).

In general, then, globular proteins have been shown to consist of alternating segments of $\alpha$-helices and $\beta$-pleated sheets, often with very irregular regions and the whole folded in a fantastic or "crazed" way. The more proteins are solved to atomic detail, the more remarkable are these structures seen to be. It was apparent that no

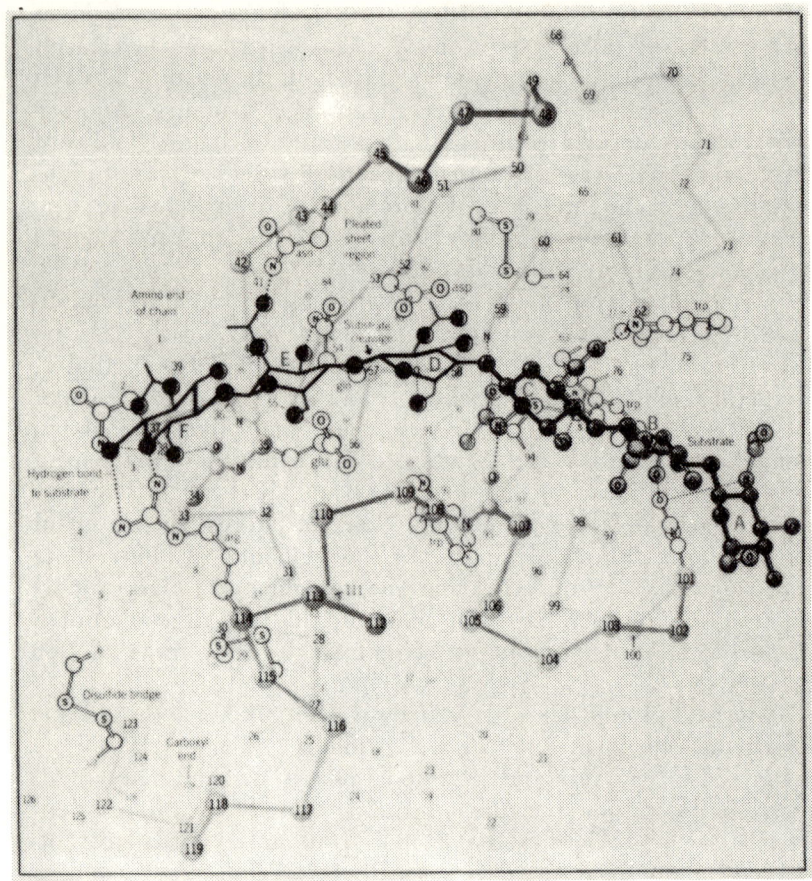

**Figure 17.** Lysozyme chain folding. This was the first enzyme to be solved (by Dr. D.C. Phillips and colleagues). The cleft that forms the active site runs horizontally across the molecule and here has a hexasaccharide subtrate bound within. Side chains that may link the substrate are shown. Reprinted with permission from *The Structure and Action of Proteins* by Richard E. Dickerson and Irving Geis. W.A. Benjamin, Inc., Menlo Park, California, Publisher. Copyright © 1969 by Dickerson and Geis.

inspired guess or model-building based on any rule or regularity of structure could have predicted their form. Since a decade ago when the solutions began to appear, scientists have concentrated their attention on trying to understand the rules for folding, and in trying to relate the activity of a particular protein to its unique structure.

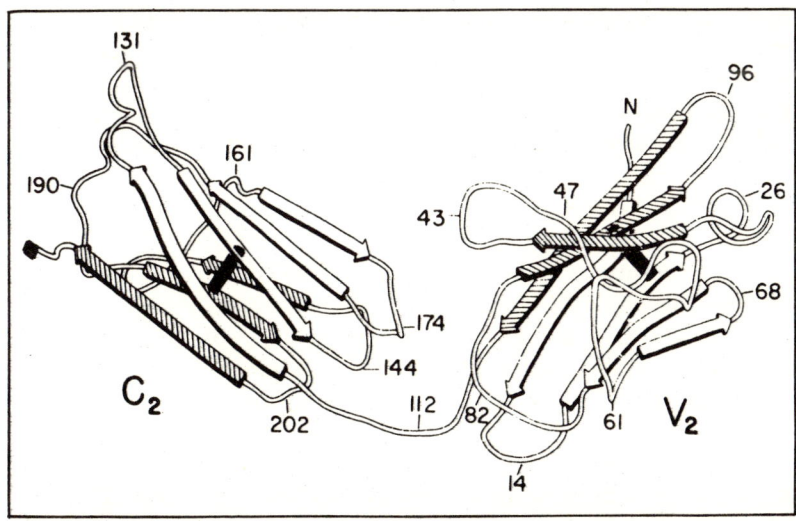

**Figure 18.** $\beta$-barrel structure. Schematic drawing of chains from certain immunoglobin domains (see Figure 21). In many globular proteins the polypeptide chains in the $\beta$ conformation wind as "staves" in a barrel to build a stable closed structure. Reprinted with permission from *Biochemistry* 12, No. 23 (1973), p. 4628, by Schiffer, M. *et al.* Copyright by the American Chemical Society.

## PROTEIN "MACHINES" HAVE MOVING PARTS

About 50 globular proteins have now been visualized in atomic detail, and it has been possible to begin to understand how many of them work. In a number of cases, small movements are involved in the functioning of the molecule. In this sense, we can view proteins as machines. The eye perceives only the final results, but x-ray crystallography and chemistry together reveal the actual workings on a much finer scale in the realm of atoms and electrons.

We have mentioned earlier that enzymes are proteins which speed up reactions that otherwise would not take place: they are catalysts. Plausible molecular mechanisms for enzymic action have been deduced in a number of cases. Here we may describe one example, which is possibly the best understood at present.

Chymotrypsin is one of the enzymes that digests protein foods in the mammalian gut. That is, it is a protein that cuts apart the peptide links in the polypeptide chains of protein foods. Chymotrypsin is secreted by the pancreas in an inactive form, which becomes active upon cleavage of one of its bonds by another digestive enzyme, trypsin. Now, chymotrypsin is one of a very large family of enzymes, including trypsin, a number of proteins crucial to the blood clotting system, and certain proteins involved in the immune reaction. All of these enzymes are proteolytic, or "protein-cutting." But they differ in specificity. Hence they are used to perform different biological roles.

The structure of chymotrypsin was determined by David Blow and his colleagues at Cambridge in 1967, and during the past decade a very convincing model for its action has been deduced.[29] It is a relatively simple protein whose chain has much $\beta$-pleated sheet structure (figure 19). A few key residues, including the amino acid serine, are essential for the enzyme's activity. If these are modified, the enzyme does not function. Moreover, these critical amino acids are widely separated in the linear sequence of the polypeptide chain of chymotrypsin, but they are brought together in the three-dimensional architecture of the protein. Another important feature of the molecule is a deep, narrow pocket which is the receptor for the specific side chain of the protein that is to be split. The shape of this pocket and its chemical nature define the specificity of the enzyme. Now, we cannot here go into the many technical details of precisely how this peptide chain is cut. Simply put, the side chain next to the bond to be cleaved is inserted into the binding pocket of the enzyme, triggering movement of a serine group on the enzyme. This causes the transfer of a proton, or hydrogen ion, along three precisely oriented amino acid residues of the enzyme, to reach the susceptible peptide bond in the protein. This bond becomes unstable and breaks. The system is then regenerated, incorporating a water molecule from the aqueous surrounding. Such a scheme, which I have vastly simplified, is called a "charge relay" system. It involves much sophisticated chemistry and a number of steps that I have omitted here. But a point which should be emphasized is that during this catalytic process, there are precise movements of side chains on the enzyme as they swing out to achieve the correct orientation for the reaction to proceed. The detailed geometry of these movements is crucial for the reaction. A similar mechanism appears to exist for the other enzymes of this family, and they are all called serine proteases, although there are various differences due to the nature of the protein acted upon and certain details of the mechanism itself.

A most interesting recent finding is that by Huber and

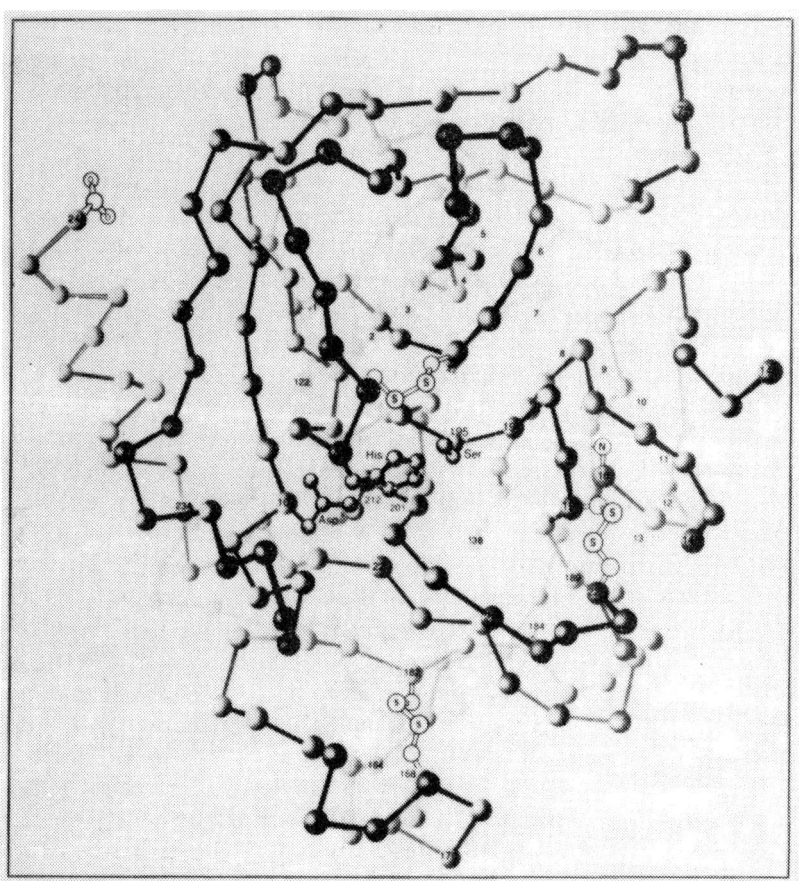

**Figure 19.** Chymotrypsin chain folding. In this illustration only the α-carbons of the polypeptide chains are shown. A "binding pocket" or crevice in the molecule accepts large, hydrophobic side chains. From Stroud, R. M., "A Family of Protein-Cutting Proteins," *Scientific American* 231 #1, 74. Illustration copyright © 1974 by R. Stroud, R.E. Dickerson and I. Geis.

collaborators,[30] who have shown by x-ray crystallography the architectural difference between the inactive and active form of the related enzyme, trypsin. It has turned out that trypsinogen, the inactive molecule, has a crucial domain that appears flexible or poorly ordered as seen by x-ray crystallography. Upon activation of the enzyme by the cleavage of one bond and the removal of a small peptide, the active enzyme trypsin takes on a relatively rigid and well-defined structure all of whose parts are seen in great detail by x-ray crystallography. In this case, flexibility of part of the molecule is a mechanism used to shut off the activity of the enzyme; that is, no precise fit with the substrate can be achieved in a chain that wobbles.

Other kinds of mechanisms are now being worked out for the great variety of enzymes whose structure has been established. It appears therefore fair to say that at this time we understand, at least in certain cases, the relationship between the three-dimensional architecture of an enzyme and the way in which it works, and that this achievement is due to the results of x-ray crystallography together with the techniques of physical chemistry.

Another protein whose mechanism is becoming understood is hemoglobin, the oxygen-carrying protein in our red blood cells. As Harvey showed, the heart can be considered a pump and the blood vessels a hydraulic system. And, on a much finer scale, the crucial protein in the blood circulation can also be conceived of as a molecular machine with moving parts. Hemoglobin is designed to carry oxygen from the lungs to the tissues and to facilitate the transport of carbon dioxide back to the lungs. Without this protein, blood could transport relatively little oxygen because of oxygen's limited solubility. As we have mentioned earlier, hemoglobin was shown by Perutz to consist of two pairs of subunits, each folded very similarly to the way the chains are folded in myoglobin. Each hemoglobin subunit carries a special flat group of atoms containing iron and called a "heme" group. This is the part of the molecule that gives both hemoglobin and myoglobin their reddish color. In hemoglobin, each of the four protein subunits carrying one heme group can combine with one oxygen molecule. The polypeptide chains of each of the subunits are highly $\alpha$-helical and form a kind of container or basket for holding the critical heme group.

Hemoglobin is a stable association of subunits, and this design, in common with a number of other multi-subunit proteins, allows its function to be closely regulated.[31] The remarkable feature of this kind of molecule is that the binding of oxygen to one heme group promotes the binding of oxygen to the other heme groups on the protein. Similarly, the release of oxygen from the molecule is enhanced once oxygen begins to unload. There is "cooperativity" among the subunits. Thus, the physiological function of respiration is carried out by a protein molecule, which appears to be exquisitely designed for this purpose. But how the information that oxygen is bound is actually transmitted to other subunits large distances away remained a mystery for a number of years.

Since the crystallographic solution of the structures of both the oxygenated and deoxygenated forms of hemoglobin by Perutz and his colleagues, intensive study of its structure in relation to the chemical reactions it can undergo appears to have given at least a good working model for this remarkable molecular machine.[32] The basic idea was advanced by Perutz in 1970.[33] He showed that there are essentially two structural arrangements for the subunits in hemoglobin: one existing in the oxygenated form, the other in the deoxygenated form (figure 20). The trigger for the change in hemoglobin upon oxygenation is a very small movement of the iron atom (about ¾ of an angstrom) into the plane of the flat heme ring. In the deoxygenated state, the radius of the iron atom is too large to fit. This movement then causes the amino acid residue histidine to move also, setting into motion a series of atomic displacements that propagate along the molecule. There are finally structural changes at the surfaces of all the subunits so that their relative positions change. Perutz has compared the movements of the hemoglobin subunits with a lung: here the slight expansion between pairs of subunits occurs in the form where oxygen is low, while taking up oxygen leads to a more compact structure. In this sense, the respiratory protein itself breathes.

Many aspects of the mechanism of hemoglobin, as well as other proteins containing subunits, are being analyzed with detailed information related to this kind of model. Moreover, many pathological states of hemoglobin are becoming understood at the atomic level. One classic case is that of sickle cell anemia. Sickling is a genetic disease resulting in a change in the shape of the red blood cells and an increased fragility. The gene causing sickling also confers resistance to malaria; it is therefore conserved in areas like Africa, where malaria is endemic. Thus, sickle cell anemia is largely found in Blacks. In 1949 Linus Pauling recognized that sickling is a molecular disease dependent on a change in the hemoglobin molecule. The chemical nature of this change was found in 1954, when Vernon Ingram showed that the substitution of a single amino acid for another had profound effects on the molecule. This substitution results in a stickiness of the deoxygenated form of the molecule, which then aggregates, changing the shape and properties of the red blood cell and leading to the clinical symptoms of the disease. A

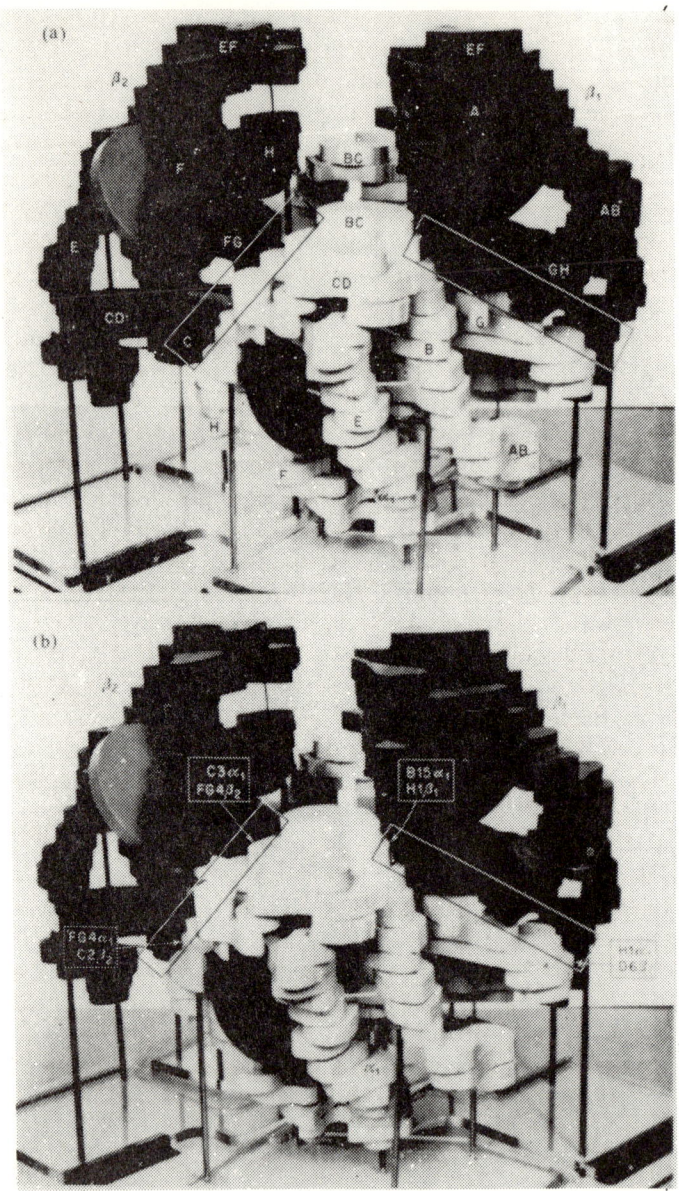

**Figure 20.** Hemoglobin molecule: (a) deoxygenated and (b) oxygenated forms. The most striking change in the structure is the increase in separation of the $\beta$ chains in the deoxygenated state. *Photograph courtesy of Dr. M.F. Perutz.*

number of other abnormal and mutant hemoglobins have also been studied by x-ray crystallography and physical methods in an effort to comprehend the underlying basis of their properties, and, perhaps, to reverse some of their undesirable effects. The substitution of a different amino acid in an important position, either related to the heme or to one of the contact regions between subunits, can have a major effect on the properties of the molecule. Hemoglobin is, therefore, one of the best examples we have where molecular biology can provide exact answers to medical questions.

The third protein whose design and function I shall describe is the antibody molecule, the basic protein of the immune system. Certain cells in the body produce antibodies in response to a particular foreign agent, called an antigen, and this results in a mixed (heterogeneous) population of proteins. In some malignancies, however, a large amount of one kind of antibody or immunoglobulin is formed. This material has been used for structural studies.

Two aspects of antibody function may be distinguished. The antibodies must recognize certain antigens by one part of their structure and eliminate them with another part of their structure. These two features are coupled in the molecule. There is thus a signal transmitted from one part of the molecule to another. The crystallographic solution to antibody molecules has come rather later because of problems of heterogeneity, but during the past five years a number of structures have been solved.[34,35,36,37] These results have given us a picture of some aspects of the molecule's function.

The complete amino acid sequence of an antibody was determined by G. Edelman and co-workers in 1968 and many of its properties were clarified. The immunoglobulin molecule is Y-shaped with a stem and two arms (figure 21). It consists of two identical pairs of chains, periodically cross-linked by disulfide bridges. The fact that the molecule is a dimer means that there are two combining sites to capture an antigen and also allows a cross-connected network to be formed. At the two antigen-binding ends, or tips, of the molecule are what are known as hypervariable regions. These define the combining specificities of the antibody. The rest of the molecule shows a striking pattern of regions that have related compositions and similar structures. These compact "domains" are about 40Å in size, and consist of polypeptide chains in a β-pleated sheet structure. A basic aspect of the molecule is that the hypervariable regions of the chain at the ends of the arms are exposed;

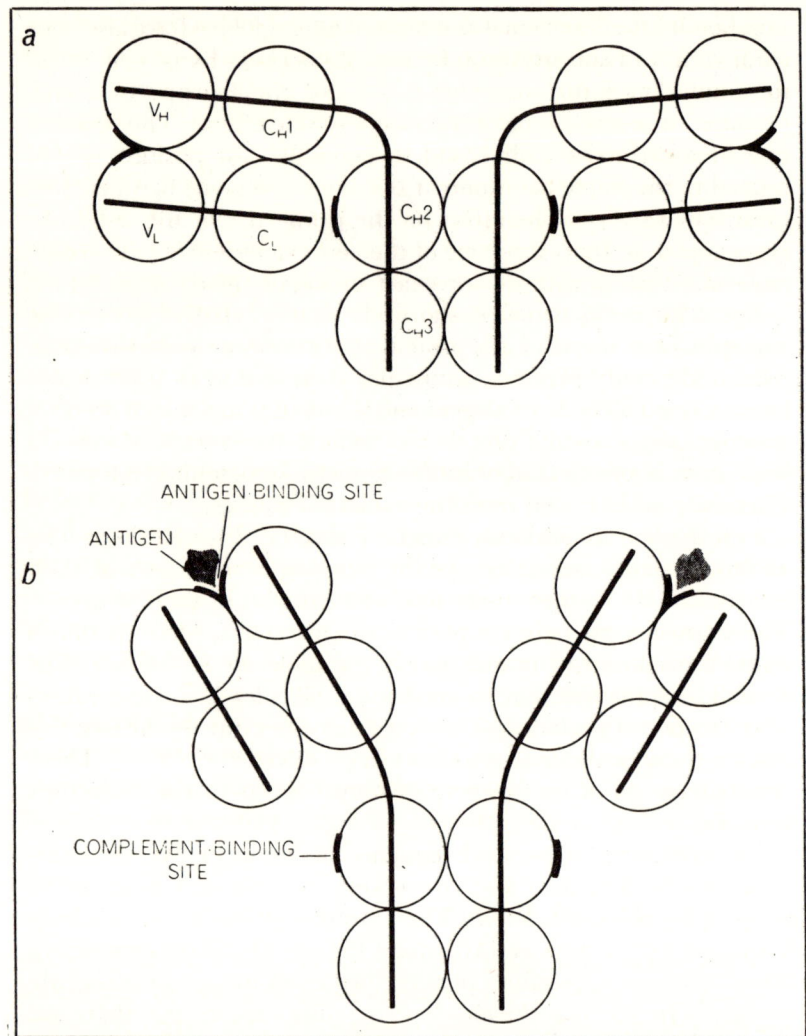

**Figure 21.** Immunoglobin molecule: two forms. Model showing the Y-shaped molecule made of compact domains with "hinge" regions. Antigen binding at the tips triggers the change in the shape of the molecule. From Edelman, Gerald M., "The Structure and Function of Antibodies," *Scientific American* 223, #2, 34. Copyright © 1970 by Scientific American, Inc. All rights reserved.

these parts of the structure face out toward the environment and hence are not constrained by contacts with other regions of the chain in the molecule. The molecule is thus designed so that there may be a great variety of substitutions, hence a great range of specificities in the combining sites.

Now, for a number of years it has been known that immunoglobulins are flexible. There are, in fact, hinge regions between the stem and two arms that appear to be the swivel points. This structural feature allows the molecule to change from a "Y" to a "T" shape, and so have the arms swing out to interact with distant regions of an antigen or adjust to different spaced sites on the antigen. Recently, Huber and his colleagues,[38] from x-ray crystallographic and chemical results, have proposed a model that identifies additional hinge regions in the arms that may relate to the cooperative behavior of the molecule. They suggest that antigen-binding causes a stiffening of the flexible antibody molecule "by formation of specific interdomain contacts. The major hinge is then essentially deleted — or undergoes a large structural change — and the molecule changes from a Y to a T shape. The stiffening of the molecule upon binding may be the signal for events leading to the elimination of foreign molecules." This model is speculative, and many areas in this field are under intensive study. But it shows the continuing power of crystallography to generate a model that accounts for function.

In the preceding examples, motions — either of side chains as in enzyme catalysis, or between subunits in regulatory proteins such as hemoglobin, or of the hinge regions and domains in antibodies — may be viewed as the essential parts of "proteins as molecular machines."

## RULES OF THE GAME

What makes these proteins fold up as they do? Clearly, there are characteristic folds of polypeptide chains common to families of proteins despite wide variations in sequence. This is seen, for example, in related classes of enzymes or in the heme-containing proteins. But what are the rules for folding? From a given linear sequence of amino acids, can we predict the three-dimensional structure of the native protein?

That all the information is present in that linear sequence has been known for some years. It has been shown that when polypeptide chains of a protein are unfolded by substances that break the weak intramolecular bonds, the randomly coiled chain can re-fold into the native structure under the correct solvent and temperature conditions. So, the information for folding is encoded in that linear sequence of the chain. And the "inner logic" of folding is known to the chain — but it is not yet known to us. This is one of the last and most formidable problems to be solved in all the riddles of protein structure.

During the past 15 years, there have been a number of attempts to solve this problem. Analyses of the energies of polypeptide chains, based on contact distances between atoms and other constraints, were early undertaken, and Ramachandran and collaborators showed that, indeed, the right-handed $\alpha$-helix and the pleated-sheet structures and the polyproline collagen-type fold were conformations that were energetically favored.[39] But given an array of different amino acids, what final fold would be taken up? A number of empirical approaches to this question have been put forward. As more protein structures are solved, the probability of a particular amino acid being found in a particular kind of fold can be estimated. Yet, it has recently been shown that the predictive value of such correlations is high only when much $\alpha$-helix is present. Proteins that have different designs or less regular regions do not yield their secrets so readily.[40,41] And these results again apply only to the first part of the question: how much helix or pleated sheet exists in various regions of the linear sequence? But even if these predictions were very good, how are these parts of the chain related in space? We wish to know not merely how much of the chain might be $\alpha$-helical or how much might be $\beta$-pleated sheet, but what would be their arrangement relative to one another. Could the three-dimensional folding be predicted?

With 70 or so high-resolution structures now available, more regularities in folding in the globular proteins are becoming discernible. The $\beta$-barrel mentioned before is one such structure, and other patterns of regularity are being detected (figure 22). One of the difficulties in any predictive scheme, however, is that how a chain folds depends not only on its secondary structure — that is, its local winding — but also on the interactions that are made with more distant portions of the chain. As more structural patterns are discerned, perhaps some useful classifications will emerge. One recent such study[42] envisages a folding picture where "pieces of secondary structure first diffuse together to form folding units and then associate to form the native structure." A search for locally ordered units seems a plausible approach to the difficult problem of compre-

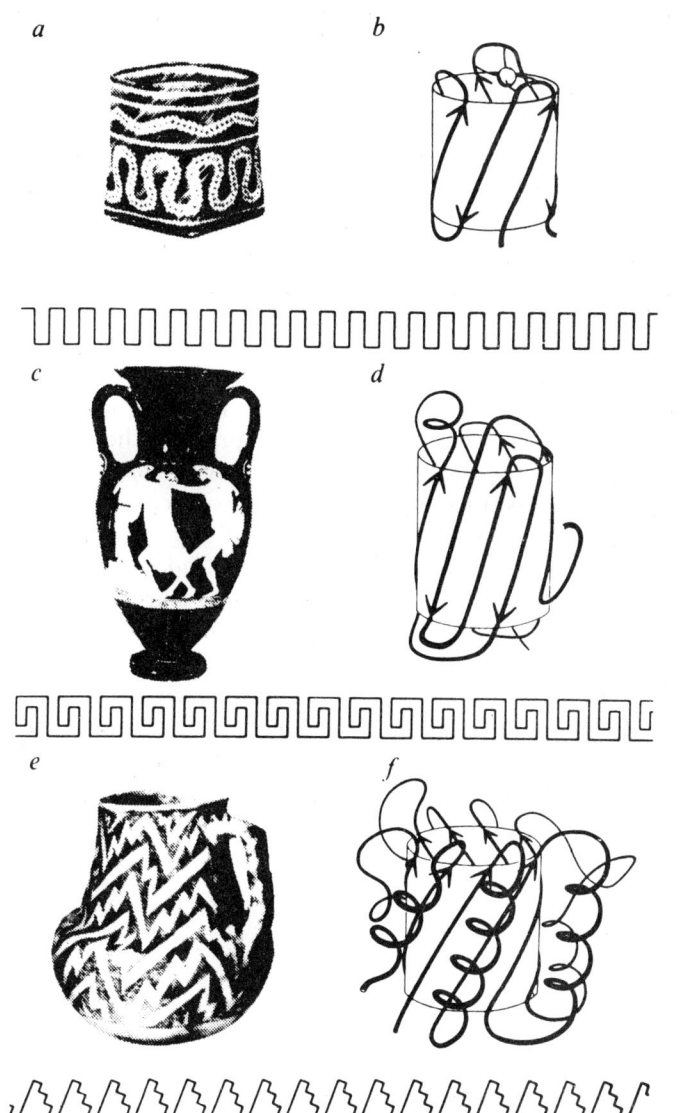

**Figure 22.** Geometric motifs in protein folds. Comparison of common Greek and American Indian geometric motifs and $\beta$-barrel structures in certain globular proteins. This illustration dramatizes the simplicity of folding present in parts of complex protein molecules, as in good pottery design. Reprinted with permission of Dr. Jane Richardson, from *Nature* **268**, 495, 1977.

hending the complex yet organized architecture of globular proteins.

## CONCLUSION

Proteins exhibit great diversity, yet many common features. As Jacob has pointed out,[43] "what characterizes the living world is both its diversity and its underlying unity." Natural selection has been operating on protein structures since they were first formed, and how they have evolved is now becoming comprehensible. "Proteins fulfilling similar functions in different organisms frequently have similar sequences, but . . . proteins with different functions often exhibit rather large segments in common." A mechanism where genes duplicate appears to account for these facts and also allows for diversity. When one copy of a protein is present, the other is released, as it were, from the constraints of natural selection and can mutate more or less freely so that changes may accumulate and new functions develop. Both hemoglobin and $\gamma$-globulin are examples of duplication from a common genetic origin, leading to very sophisticated molecules. But as Jacob has stressed, evolution is a kind of molecular tinkering: "It is always matter of using the same elements, of adjusting them of altering them here and there. It is always a matter of tinkering." As we learn how proteins are made, so will we understand better how proteins have evolved. We therefore seek to know not only how proteins work the way they do, but how they came to be that way.

# THE STORY OF DNA

## Ruth Hubbard

*Dedicated to the memory of Dorothy Wrinch
and of her daughter, Pamela*

*I shall depart from the usual academic style and begin with a brief excursion into autobiography, because I would like to convey some of the many ways in which I think and feel about the subject. Objectivization and the resulting reification of feelings and relationships is a severe limitation of the scientific method and of the way in which its results are formulated and communicated as science. And I firmly believe that we need to come up with an analytic approach that does not reject the tools of science, but that is sensitive to the role of our subjective participation in the science we construct. For the "data" and "facts" of science, as of all knowledge, lie in our sense experiences (and in their technological extensions through machines). Therefore they depend on where we seek, what we attend to, and how we organize it in our minds. Hence the admission and description of our personal involvement with our subject must be part of the picture.*

## 1. Autobiographical

I met Dorothy Wrinch and her young daughter in 1947, the first summer I spent at the Marine Biological Laboratory in Woods Hole. I was a beginning graduate student at the time and was assisting in a section of the Physiology Course; Dr. Wrinch, much older, was teaching another of the sections. It is a very aesthetic memory, since she and her students always were surrounded by interestingly shaped objects, that they were constructing from different paper cut-outs in order to illustrate and describe the shapes and symmetries of crystals. Their work was very clear and clean; and all the while I was wading in salt water, killing and dissecting animals, extracting their tissues, and measuring absorption spectra. We were looking at the world through differently colored spectacles and coming up with different kinds of pictures that require the transformational logic of science to bring them into the same universe. But I wasn't thinking about that then.

A year later, I lived in London, the post-World War II London of rationing, queuing, bomb-sites; of a spirit of victory over bombs and missiles (familiarly known as doodle-bugs); of a labor government that had been elected by a strong majority, and was introducing a National Health Service to become a model for the rest of the world. It was a drab and dowdy London, drizzly and foggy and cold; a happy and optimistic London, full of well-fed and cared for children, the country's special pride and concern; the London I fell in love with, and the London a young English chemist, named Rosalind Franklin, gladly exchanged for Paris — at least for a time.

I worked there for more than a year, then came back to the United States for my Ph.D. But in 1950, I was back in Europe for the summer, this time at the Carlsberg Laboratory in Copenhagen, where the Carlsberg brewery rationed each of us to only six bottles of beer (or soda water) a day. When I visited the laboratory of the biochemist, Hermann Kalckar, before returning to America in the fall, I found him all excited because two bright young phage workers were about to come to him from Max Delbrück's laboratory at Cal. Tech.: Gunther Stent and Jim Watson. Watson has written about his Copenhagen period in *The Double Helix*.[1]

When I returned to Copenhagen in 1952, on fellowship for a year, Watson had moved to Cambridge, England. He hadn't accomplished much in Copenhagen other than accompany Kalckar and others to Naples for the spring. What he was doing in Cambridge we found out in April 1953, when he and Francis Crick published their immediately famous paper in *Nature* that proposed the double helical model of DNA.[2] Five weeks after the appearance of this paper, on that rainy June day in 1953, when the London streets were a dense crowd of people, come to see Elizabeth crowned Queen of England, I found myself quite by accident in the Cavendish Laboratory in Cambridge, listening to Francis Crick expound on their exciting new DNA structure. I did not meet Watson until he came to Harvard a year or two later. We became friends, and remained friends until quite recently, when disagreements over the story of Rosalind Franklin and the controversy surrounding the recombinant DNA technology stretched our relationship beyond the breaking point.

My years of scientific apprenticeship and intercontinental wandering therefore overlap Watson's. We spent our scientific adolescence at the same time and in similar places. When I read *The Double Helix*, I recognized attitudes, friends, and sights. When I first read it in manuscript, I liked it tremendously. I felt that it described scientific work "like it is." I knew nothing of Rosalind Franklin, had not met her or even heard of her (to remember, that is), and such was my state of integration into male-dominated science that I noticed nothing wrong. (Let me say at once that many others, who also didn't know Franklin, were a good deal more alert than I to the outrageous sexism of Watson's descriptions of her as a woman and as a scientist.)

I first became aware of the scandal surrounding the Watson-Crick model of the double helix when a friend, herself a crystallographer, gave me the manuscript of Anne Sayre's book about Rosalind Franklin.[3] I read it and was appalled. I have since checked many of Sayre's facts and spoken with a number of Franklin's friends and acquaintances. I have summarized my impressions in a review I wrote of Anne Sayre's book and will recapitulate some of them here.[4] But this brings me back to Dorothy Wrinch.

During the last few years of Dorothy's life, I used to visit her often at her home in Woods Hole. And on one of the last of these occasions, I brought her a copy of Anne Sayre's book and of my review of it, and we talked at some length about Rosalind Franklin and about some of the problems women face in science. It is, therefore, appropriate for me to write in Dorothy Wrinch's memory about the story of DNA.

## 2. Philosophical

The art historian, Ernst Gombrich, has quoted one of his colleagues as saying "that all pictures owe more to other pictures than they do to nature."[5] And so it is with science. Nature, of course, is in there; but the primary linkages are with the science that has gone before. In a way that is only to say that like art, science is a product of the human imagination, and like art, to be acceptable and have meaning, it has to be comprehensible in terms of our experience of the real world and of the accepted interpretations of it. When the imagination becomes free-wheeling and loses all linkages with the real world, or rather, with the world that we accept as real, we call it diseased or insane. But in the activity of making science (and science of course is *made;* it isn't "out there") we abstract from what Alan Watts has aptly referred to as the "seamless unity of nature" those things that we notice and therefore pull out of the continuum.

"Things," he has said, "are the measuring units of thought just as pounds are the measuring units of weighing."[5] And just as the enumeration of the pounds in a rock or the inches in a log are only a one-dimensional description of these objects, so the "things" or "facts" we use to signify and describe our experience of nature are but discrete fragments of its total reality. When we translate these units of thought and experience — facts and things — into words, we delimit and trim them even further; because, as Watts points out, "The fundamental realities are the relations . . . in which facts [or things] are the terms or limits. . ." So, if the world consists of an infinitely intertwining web of relationships, which our sensations and thought processes sort into things and facts, and which we delimit still further by the process of naming, science raises these stages of abstraction (or, more correctly, of concretization or reification) yet another notch. For a further selection is made when scientists turn "facts of nature" into "facts of science."

Scientists don't just hold up a mirror to nature. They use something more in the nature of a coarse sieve through which drop all the things that they either don't notice or those that they take to be irrelevant to the inquiry at hand. The real intellectual labor of scientists consists in constructing a coherent picture of the world from what they sift out of it as noteworthy and significant. Anthropologists use two interesting terms for the process that they engage in when they enter a foreign culture and try to formulate its laws: *backgrounding* and *foregrounding*. By these words, they denote the often unconscious activity of allowing certain things and habits and events to merge with the background of the unnoticed or the "unnoteworthy," while pulling others into the foreground where they command attention and have to be dealt with. And I would argue that a similar process occurs in the natural sciences. Scientists make juxtapositions and construct relationships out of the "facts" into which they sort or delimit the complex web of interdigitating relationships that constitutes the real world. To extend Watts's image, we can say that scientists transform the seamless unity of nature into a carefully patterned patchwork quilt. And it is important to be aware of the elements of both patchwork and patterning: for both involve choices that are far from arbitrary.

But the way in which the patches are selected — the unconscious decisions on what remains background and what gets pulled into the foreground — and the way in which they are stitched together, are determined only in part by the explicit postulates and rules of the "scientific method." For they, as well as our total selective mind-set are social products that depend on who we are and where and when. Our scientific reality, then, like all reality, is a social construct — by which I don't mean to say that I am an 18th century idealist and believe that there is nothing out there. I am quite sure that there is something out there, but I believe that what we see out there and our interpretation of it, depend on the larger social context that determines what we actively notice and accept as real.

We learn to see the world while we are infants and young children, and we do it by a process that requires patient guidance from all the adults around us. By the time we are educated, which literally means "led out of" or "away from" (from whom? our selves?), we end up so transformed that we have no way of remembering how we saw the world when it truly was our own, when we were tiny children and hadn't yet been taught what the world is "really" like. In fact, the very act of language acquisition in young children in part is precisely this kind of teaching of the important social, and therefore verbal, distinctions and categories. When a child has been corrected often enough that "this is a doggy and not a kitty," s/he has not only learned two words, but s/he has also learned that there are two different kinds of domestic animals with fur and a tail, and that it is important for an educated, grown-up person to know them apart. And when a child learns much subtler distinctions, such as "no, dear, this is a boy, not a girl", the very subtlety, coupled with the insistence that the distinction be made correctly and every time, convey a profound social message. In another culture or in another natural setting, this kind of emphasis may be put on different textures of snow or on the different sounds made by water. It just depends on what is important in life and in one's particular setting.

What I am trying to say is that the most profound and significant concepts are often internalized quite without our knowing it, and usually only people who are, for some reason, marginal to the prevailing belief system — radicals of some sort, or women — are even tempted to become conscious of the process and to question the results. That, of course, is what "consciousness raising" is about.

## 3. Sociological

The story I am focussing on has to do with styles of work in science, with what one considers important in it, and with the ways in which women and men live in scientific laboratories.

The first question I want to ask is: Why was DNA "the most golden of all molecules,"[7] the most important molecule in biology — the molecule that a bright and ambitious young biologist, a former Quiz Kid, named James Dewey Watson, might reasonably decide at age 19 to be the key to understanding what life is, as well as, perhaps not incidentally, to his own success? For Watson made a very conscious choice.[8]

I would like to suggest that perhaps a less individualistic society, one that cared less about the traits, accomplishments and successes of *individuals,* might also care less about how these are passed on from parents to their very own, individual offsprings. Such a society and the people in it might find many aspects of biology more interesting than heredity, genes, and hence (in a reductionist science) ultimately DNA. To appreciate this point it is useful to read the literature of the early and mid-nineteenth century, when there was no proper genetic theory — such as the novels of Dickens or George Eliot and the scientific writings of Darwin's cousin, Francis Galton. In a book entitled, *Hereditary Genius,* for example, Galton sets himself the task of understanding why it is that as one reads down the roster of the best students at Cambridge University, one keeps coming up with the same surnames in generation after generation. And the answer he gives is that the factors that make for these successes must be inherited biologically. (However, he fails to notice that they are inherited only in the male line.)

It seems clear that, beginning with the period of the American and French Revolutions, the meritocratic philosophy of Euro-American liberalism, combined with the unwillingness of the ruling classes to share their power, made it mandatory that the men from the educated, upper classes find biological "reasons" why they were ruling the world.

That is one point. But I would also suggest that even in a highly competitive and individualistic society, we would never expect that we could answer the question, what is life, by studying genes, had we not accepted Descartes' definition of organism as machine. It is this that leads us to expect to answer such questions as what life is, by taking living organisms apart into smaller and smaller units — to reduce them to chemistry and physics — even though the attributes of "life" or "living" be lost in the process. Nor might we expect to learn how genes work from studying molecular genetics, the branch of biochemistry that its founders have called "molecular biology," so implying in the very name that DNA is the only important molecule in biology.

I emphasize this because Watson has written that he and Francis Crick, quite separately and long before they knew each other, were led to their keen interest in genes and DNA by reading Schrodinger's *What is life?*[9] This little book, published by one of the great physicists toward the end of World War II, drew the attention of many physicists to biology. For at the war's end many physicists were disappointed with what had become of the physics that at the turn of the century had produced the far-reaching generalizations of relativity, quantum theory, complementarity, uncertainty. By 1945, physicists had generated two atom bombs that had been dropped on people. Furthermore intellectually, at least in the view of some physicists, physics was beginning to degenerate into a queuing up in front of bigger and bigger machines so as to produce smaller and smaller particles. Many bright physicists were looking for more interesting and meaningful problems and some of them were excited by Schrodinger's promise of the gene as the new frontier.

It is worth realizing that in the story of the double helix, the only main actor who had been trained as a biologist was Jim Watson. The others were physicists or chemists. Moreover, Max Delbrück, on whose ideas Schrödinger based much of his discussion of the gene, was promising not only that life would ultimately be understood by reducing it to physics (not chemistry: Delbrück didn't like chemistry), but that in the process new physical laws were likely to emerge. This was one of the attractions that brought Delbrück himself in, and it may have brought in some of the other physicists as well. Watson has written that one reason he decided to work for his Ph.D. with Salvador Luria was because Luria had worked with Delbrück and they were close friends.[9] And indeed his association with Luria quickly brought him into contact with Delbrück.

In the event, learning the structure of DNA has brought no new physical laws, nor do we know what life "is." But we have learned a lot about the biochemistry and structure of those big molecules, the proteins and nucleic acids, that occur by and large only in living organisms. If we ask further, whether the elucidation of the structure of the double helix has introduced a new paradigm into biology and produced a revolution in our way of thinking about organisms, I would say that it has not. On the contrary, the discovery of the double helix and everything that has since happened in molecular biology, lie squarely within the dual paradigm of organism (1) as machine, and (2) as a machine that will be understood better and better as we come to understand the workings of its smaller and smaller parts.

Before getting into the scientific part of the story in more detail, it is worth asking what we can learn from Watson's *Double Helix* about the way science is done, and particularly, about the roles women and men play in the scientific enterprise. This is not a trivial question because the book *was* a best seller and still is required reading for many high school students. So let us look at the way he introduces the four principal characters in his story: himself, Francis Crick, Maurice Wilkins and Rosalind Franklin. (Watson also lists Linus Pauling as a main player, but he comes in quite late in the book and is never rounded out like the others, who enter the scene in the first half dozen pages.)

We meet Watson already in the introduction, climbing in the Swiss Alps with a fellow scientist, and being hailed by a scientific acquaintance from London as "Honest Jim," the phrase he originally intended to use as the title for his book. (It is my memory that he didn't because of the similarity to the title of Kingsley Amis's *Lucky Jim*.) Crick enters in the first sentence ("I have never seen Francis Crick in a modest mood. . .") and by the fourth, is likened to Rutherford and Bohr. For Wilkins we have to wait till the second chapter (only three pages further on) where we learn that "At this time [fall of 1951] molecular work on DNA in England was, for all practical purposes, the personal property of Maurice Wilkins, a bachelor who worked in London at King's College. Like Francis [Crick], Maurice had been a physicist and also used x-ray diffraction as his principal tool of research. . ."

So at once the stage is set with our three male principals: Watson a bit frivolous and problematic (though clearly extraordinary: after all he is the one who is writing); Crick loud, ebullient, and brilliant; Wilkins serious and a bit musty. The latter qualities are emphasized within another paragraph by the statement that "Maurice continually frustrated Francis by never seeming enthusiastic enough about DNA." At the end of this paragraph Franklin makes her entrance: "Moreover it was increasingly difficult to take Maurice's mind off his assistant, Rosalind Franklin." And we must pause at once to understand that Franklin was *not* Wilkins's "assistant;" that their appointments were independent, equivalent and both in the laboratory headed by John (later, Sir John) Randall, the professor at King's. Furthermore, though Franklin was a newcomer to this laboratory in which Wilkins had been for some time, she was invited to come because she knew *more* than Wilkins about x-ray diffraction, and had been led to believe that she would be working there on her own. But let us go on with Watson's description:

> Not that he was at all in love with Rosy, as we called her from a distance. Just the opposite... Maurice, a beginner in x-ray diffraction work, *wanted some professional help* and hoped that Rosy, a trained crystallographer [incidentally, she was not that], could speed up *his* research. Rosy, however, did not see the situation this way. She claimed that she had been given DNA for her own problem and would not think of herself as Maurice's assistant. (My parenthesis and italics)

Then follows a discussion of her lack of femininity and her unattractiveness, her hair, clothes and grooming. It ends with the statement that "Clearly Rosy had to go or *be put in her place*. The former was obviously preferable because, given her belligerent moods, it would be very difficult for Maurice to *maintain a dominant position that would allow him to think unhindered* about DNA." (My italics)

These passages speak for themselves. The only factual correction that needs to be made is Sayre's insistence that no one — neither family nor friends — ever addressed Franklin as Rosy; that this is part of the stage-set in which this dowdy (she happened not to be that), petulant, uppity blue-stocking claimed as her own, work that rightfully belonged to a serious, albeit somewhat stodgy, and of course male, scientist. As the story goes on, this picture is embroidered and amplified. And what we learn from the entire book is that real science — good science — is something that is done by men, indeed very bright and ambitious men, who relate to women in two ways: if they are beautiful and charming, women offer a delightful escape from important and serious concerns, such as finding the key to what life is. But women can also be a damned nuisance; particularly if they aren't sufficiently helpful and submissive technicians and, worse yet, if they try to be scientists and follow their own ideas. In that case, they can stem progress and even stop science dead.

I want to digress for a moment because I think it is worth asking whether Watson is unique in this aspect of his description of science. And for comparison, I want to look at two earlier stories that describe science and scientists: Sinclair Lewis's *Arrowsmith* and C.P. Snow's *The Search*. Written respectively in the 1920's and 30's, these books are very different from *The Double Helix* in that they picture scientists as highly idealistic and dedicated to the search for Knowledge and Truth, and with much less interest in worldly honors and success than does Watson. The protagonists have an eye much less on the main chance than has Honest Jim. In fact, when the hero of C.P. Snow's novel recognizes that he has begun to do his science for the sake of the status and worldly success that it will bring him, he promptly decides that he'd better get out and quick. And Arrowsmith, of course, ends up doing his research in a small laboratory off in the woods and away from worldly distractions.

But both books agree with Watson's description of the roles of women. In them, too, women exist to nurture and sustain the hard-working scientist and to offer *him* (and I don't ever use male pronouns generically) much needed relief from *his* serious work. In that capacity women have major roles in both novels and are well-rounded characters, significant and strong. When they leave or die, they are sorely missed by the male scientist-heroes. But the three times in *Arrowsmith* that Lewis mentions professional women, it is to scoff. Once, while Arrowsmith is still in medical school, female colleagues are mentioned, but only as "virginal and unhappy coeds". Another time they are "emotional and frightened;" the third, they "shudder" as the male professor, the great scientist in the book, injects a virulent strain of bacteria into a guinea pig.

C.P. Snow is even more interesting, for there are several strong, important women in his book. But if a visitor from outer space were to read *The Search* there is only one point at which she/he/it could discover that women also can be scientists. Not that there is a female scientist in the book; but at one point, the hero counsels one of his more frivolous friends, for whom seducing women is a significant avocation, but who has now decided to make it in academic science. He warns him that the one thing he musn't do is to go around seducing professors' wives. His friend retorts gratefully, "I promise, Arthur, not a scientist's wife. Not one. Not even a woman scientist." And that's the only mention of women scientists in a lengthy novel about British x-ray crystallographers, a group that is unusual among scientists for counting many distinguished females among its members.

So we are in the customary bind, where when women actively participate in a field that is not stereotypically female — and not in their accepted roles as cleaning women, technicians, secretaries and wives — they are either maligned, or ignored and written out of the story.

## 4. Scientific

I shall now try to evaluate Rosalind Franklin's contribution to the elucidation of the structure of the double helix. I am using as sources things that she and others have published. Listings of the main references are available in Robert Olby's *The Path to the Double Helix*,[10] in Anne Sayre's book and in my review of it.

I shall begin in 1951 when Wilkins had been given a sample of DNA with which he and a graduate student, named R.G. Gosling, had taken a first, quite good x-ray diffraction picture. He showed this to Watson when they met in Naples that spring. Watson got excited. And since he was not enthusiastic about his situation in Copenhagen, he decided to move to Cambridge, England and learn something about x-ray crystallography in order to work on the structure of DNA. There, in the fall of 1951 he met Francis Crick who, though already 35, was working for his Ph.D. at the Cavendish Laboratory. (Watson was 23 and already had his doctorate.) Crick and Watson have recalled that as soon as they started talking together, they were struck by the similarity in each other's ways of thinking about biology. Both believed the structure of DNA to be the most important thing to work on and wanted to do something about it. Unfortunately Crick, who knew about x-ray crystallography, was still involved with the work for his thesis. So Watson had to learn quickly.

Wilkins, as we have seen, at this time was working on DNA at King's in London. His chief, Professor Randall, had invited Rosalind Franklin, who was an expert in x-ray analysis, to come and build up the x-ray diffraction unit there. Franklin had spent the preceding years in Paris doing the x-ray analysis of three-dimensional forms of carbon and wanted to turn her attention to biological substances. She therefore was glad to accept a research fellowship in Randall's biophysics unit and came there in January, 1951.

Wilkins was away when Franklin arrived and her work was underway by the time he returned. What each of them thought their official relationship to the other was meant to be is not clear; but Sayre writes that they took an instant dislike to each other and never were able to get along. Perhaps this is the time to ask whether Franklin or Wilkins were just plain difficult. Watson would have us believe that Franklin was. My own inquiries suggest that she was argumentative and at times overdefensive and prickly; but she established successful collaborations with her colleagues in Paris, with Gosling at King's, and after she left King's with Aron Klug and others at Birkbeck College in London. Similarly Wilkins has collaborated with many people. So Sayre is probably right in believing that this was an unfortunate collision of two very different, and apparently incompatible, temperaments. No doubt it was also exacerbated by antiquated sexist practices at King's, that excluded Franklin from the easy relaxed sociabilities of morning coffee and afternoon tea with her male colleagues, since they were served in separate and unequal combination rooms. (Why, we ask with Virginia Woolf, do the women always get boiled beef and prunes, while the men feast on partridges and wine?) But whatever the reasons, almost immediately tension ran high between Franklin and Wilkins.

Franklin began by building a high resolution x-ray camera to use for studying the structure of DNA with Gosling. Wilkins worked on DNA with A.R. Stokes, and there apparently was little communication between the two pairs, even though they were housed in the same laboratory. Within a few months of her arrival, Franklin discovered that, depending on the water content, DNA forms two kinds of fibers with quite different x-ray diffraction patterns. She called the dry form **A**, and the wet form **B**, and decided to begin her structural analysis by working up the **A** pattern because it showed more detail, and to go on to the **B** pattern in due course. In retrospect, this proved to be a misjudgment, but one that made sense at the time.

In November, 1951, when Watson had been about six weeks in Cambridge and as yet knew very little about x-ray diffraction, he went to King's to hear a seminar in which Franklin talked about her work on DNA. In *The Double Helix* he tells us that he did not take notes, speculated about Franklin's looks, misunderstood or misremembered a good deal of what she said and communicated his recollections to Crick the next day. Crick got intrigued, started making some calculations, and on this basis the two of them quickly built their first model of DNA. They promptly invited the group from King's to come and see it. Rosalind Franklin immediately pointed out that the data she had presented made the model highly unlikely and saw no reason to involve herself further with such idle speculations.

This encounter had a variety of important consequences. For one thing, Sir Lawrence Bragg, who headed the Cavendish Laboratory, instructed Watson and Crick not to work any further on DNA, since the King's group was working on it. But there were also subtler consequences for Watson and Franklin. Watson had been shown up by a woman, and one whom he liked to think of as Wilkins's "technician." Franklin, on the other hand, seems to have decided that Watson was a clown and to stop taking him seriously. After all he had sat through her talk and then gone off and built a model that was highly speculative and almost certainly nonsense. That was not her idea of how one did science. She saw no reason to depart from her projected course of x-ray analysis in order to engage in similar games of model building.

Watson and Crick, on the other hand, were obsessed by the realization that Linus Pauling and his colleagues at the California Institute of Technology had recently cracked the structure of proteins by the careful and imaginative construction of three-dimensional models. However, there is an important difference between the Pauling group's model building and Watson's and Crick's: prior to building their models, the group at Cal Tech had systematically investigated the structures of all the component building blocks and had established a large body of information regarding their possible configurations, bond lengths and bond angles. Watson and Crick, on the other hand, hardly knew what some of the units looked like out of which they were trying to build DNA. In the summer of 1952, Erwin Chargaff, one of the pioneers in nucleic acid chemistry, found them ignorant even of the chemical formulas of the so-called "purine" and "pyrimidine bases" that compose DNA. And some six months later Jerry Donohue, another chemist, pointed out to them that these bases can assume two different forms, of which one is considerably more likely than the other.

Late in '52, Watson and Crick heard rumors that Pauling thought he had deciphered the structure of DNA and was writing a paper about it. Through Pauling's son, Peter, they got a copy of the manuscript, quickly noticed that Pauling had made a rather basic mistake, and with that, decided that all bets were off. If Pauling was going after DNA, then it no longer "belonged" to King's; so why shouldn't they work on it, too?

It was about this time that Wilkins showed Watson Franklin's best x-ray diffraction picture of the wet (**B**) form of DNA, which clearly showed it to be helical. Meanwhile their Cambridge colleague, Max Perutz, gave Crick and Watson a copy of a privileged report to the Medical Research Council, in which Franklin had summarized her most recent findings including the spacings of the critical reflections on the x-ray diffraction pattern of the DNA fibers. Any model of DNA had to account for these spacings and for their relative intensities. So, seeing themselves in a race against Pauling and with Franklin's new data against which to check plausible models, Watson and Crick went into a frenzy of model building. Within a few weeks they arrived at the now famous structure, wrote it up, and sent the manuscript to Wilkins.

This was the first indication Wilkins had that Watson and Crick had gone back to working on DNA. Both he and Franklin immediately accepted the essential correctness of the structure, and on April 25, 1953 there appeared in *Nature* the dramatic triad of papers. The first, by Watson and Crick, described the structure; the second and third, by Wilkins and Stokes, and by Franklin and Gosling, offered supporting x-ray evidence. The papers made a big splash and Watson and Crick were instantly famous.

Franklin never knew that the crucial features of her "supporting" evidence were in their hands before they began to construct their model. At the time, she was in the process of moving from King's, which she hated, to Bernal's laboratory at Birkbeck College. There she remained, doing distinguished work on the structure of tobacco mosaic virus, until virtually the day she died of cancer in 1958, at the age of 37.

Watson, Crick and Wilkins shared the Nobel Prize in 1962; and the matter probably would have ended there had Watson not published *The Double Helix* in 1968. Until then no one knew the precise sequence of events, and hence the role that Franklin's data — unbeknownst to her — had played in guiding Watson's and Crick's formulation of the DNA structure.

Before continuing with our story, it will help to summarize the main features of the Watson-Crick model. The model pictures DNA as two ribbons, each of which consists of a long, invariant, alternating sequence of sugar (deoxyribose) and phosphate molecules: -sugar-phosphate-sugar-phosphate-. The ribbons have a polarity that allows one to tell one end from the other. Two of them, running in opposite directions (i.e., head to tail), are wound into a double helix. The two ribbons are connected at regular intervals by horizontal rungs, each of which is formed by a pair of flat discs (the purine and pyrimidine bases). There are four different bases in DNA: adenine (A), guanine (G), thymine (T) and cytosine (C). Two of them are large (the purines, A and G), and two smaller (the pyrimidines, T and C). To fit properly inside the helix, A must pair with

T, and G with C. This feature of the model incorporates observations of Erwin Chargaff and his colleagues at Columbia University, who showed several years earlier that DNAs from different organisms contain different proportions of the bases, but that in all DNAs the amounts of A and T always are equal, as are the amounts of G and C.

Watson likens the DNA structure to "a spiral staircase with the base pairs forming the steps." The helix can be very long and there can be many thousands of base pairs in a DNA molecule. In this simple, elegant structure the two base pairs therefore can be arranged in an almost infinite number of different sequences and can form a very large number of different DNAs. (This wide diversity is an essential biological requirement if the millions of different genes that code for millions of different characteristics of all the different kinds of organisms are simply different variants of DNA sequences.) That this diversity can be drawn out of such a simple basic structure is one of the features that won the model its immediate and wide acceptance.

The *Nature* paper specified all the critical dimensions, such as helical pitch, the diameter of the helix, and the number of turns between repeats along the helix. All these were obtained from Franklin's data, though this is nowhere acknowledged. Furthermore, it was Franklin who had insisted as early as 1951 that the sugar-phosphate backbone had to be on the outside of the fiber with the bases pointing inward, since that was the most reasonable way to explain why DNA fibers readily take up water. She had also suggested that the bases were probably held together by hydrogen bonds. Both these features became part of the Watson-Crick model, though this also is not acknowledged in the *Nature* paper. In a long, follow-up paper that Crick and Watson sent to the *Proceedings of the Royal Society* of London in the latter part of the summer, they wrote: "We have only considered such structures as would fit the preliminary x-ray data of Wilkins, Franklin and their co-workers. Our search has so far yielded only one suitable structure."[11] So here they say it, but this fuller account is rarely referred to or read; and it, too, does not state that they had had access to Franklin's privileged report, including her calculations. Their achievement was considerable, no matter what. Though *all* the data they used had been the work of others (Franklin, Wilkins, Chargaff, plus Donohue's suggestion about base structures), Crick and Watson were the ones who saw how to fit them together and come up with the structure. They would have been no worse off had they acknowledged this.

It is therefore interesting to see what they did acknowledge. The *Nature* paper says:

> We are much indebted to Dr. Jerry Donohue for constant advice and criticism, especially on interatomic distances. We have also been stimulated by a knowledge of the *general nature* of the unpublished experimental results and ideas of Dr. M.F.H. Wilkins, Dr. R.E. Franklin and their co-workers at King's College, London. (my italics)

The *Royal Society* paper ends:

> We are most indebted to Dr. M.F.H. Wilkins both for informing us of unpublished experimental observations and for the benefit of numerous discussions. We are also grateful to Dr. J. Donohue for constant advice on the problems of tautomerism and van der Waals contacts, and to Professor A.R. Todd, F.R.S., for advice on chemical matters, and for allowing us access to unpublished work.

There follows a remarkable final paragraph in which "One of us (J.D.W.) wishes in addition to acknowledge" respectively "hospitality", "encouragement" and general indebtedness to Sir Lawrence Bragg, J.C. Kendrew, M.F. Perutz and S.E. Luria — all of them important and powerful men, and then or future Nobel Laureates. But not a word about Franklin.

## 5. Some Final Thoughts

I want to repeat that I don't think that Watson's treatment of Franklin in the book is idiosyncratic. Perhaps a bit exaggerated; but not that unusual. Seeing a woman work in a laboratory, he is not alone in assuming that she must be somebody's technician. People walking into my office or calling on the phone, often ask me for Professor Hubbard, confident in the assumption that the Professor must be male. Furthermore, as late as 1970, so two years after the publication of *The Double Helix*, Linus Pauling, who has good reasons of his own to be leery of the scurrilous innuendos in the book, refers to Franklin's work as "Wilkins' x-ray photographs. . ."[12] Indeed as I write this I see in the *New York Times* (April 9, 1978) an article about last year's Nobel Laureate, Rosalyn Yalow, that repeatedly suggests that in the work for which she won the Prize the real, imaginative thinking was done by her male colleague, Berson, while Yalow shook the test tubes.

A question that is often asked is how "close" Franklin was to solving the DNA structure. Obviously there is no good answer. She was proceeding systematically and making good progress. She was intent on exhausting the possibilities of traditional x-ray analysis before getting involved in model building, which seems to have appeared to her a highly speculative approach. There is little doubt that had she been able to continue working on DNA, she could have solved the structure. But Franklin had decided to move to Birkbeck, because she hated her situation at King's and Sayre tells us that this had necessitated a promise "to stop thinking about DNA entirely" (to which she commented quite appropriately, "But how could I stop thinking?"). So probably it is historically correct to say — if such speculation can be called history — that she would not have solved the structure. Whether and how much the fact that Watson was breathing down her and Wilkins' necks exacerbated the situation at King's no one will ever know.

We must be clear about the fact that at the point at which Watson and Crick began their second (and successful) bout of model building, a number of important things had been learned about DNA: (1) Wilkins and Stokes, and Franklin and Gosling had shown that it was helical (at least in the B form); (2) Franklin's conclusion that the sugar-phosphate backbone lies on the outside meant that the bases must be fitted into a regular, repeating pattern *inside* the helix; (3) her x-ray data supplied the critical parameters against which to test possible models; (4) Chargaff's rules regarding the equivalence of A and T and of G and C set further limits. With all these constraints to guide the search, the structure was bound to be solved before long.

It is sometimes said that Watson's and Crick's race for the double helix and its celebration in Watson's best-selling book have lowered the moral tone of science and escalated its frenetic competitiveness, particularly in "molecular biology." I am skeptical. As I see it, science reached its *man*hood during the heyday of industrial capitalism when competition was hailed as the road to success in a system that was claimed to be meritocratic. Western scientists operate with the explicit assumption that competition sorts the chaff from the wheat, and that genuine ability is what determines competitive success, not accidents of birth, entrepreneurial skills and/or ruthlessness. During the twenty-five years since the discovery of the double helix, the number of young people entering science has increased greatly, but the number of available top positions has not. I attribute the deterioration in the social relationships and mores — the enormous competitiveness and secretiveness that poison the contemporary scene for many aspiring young (and no so young) scientists — on this larger social context more than on the "winner takes all" morality, exemplified in Watson's race for the Nobel; which is not to deny that the overwhelming success of this race and the power and prestige it quickly brought the till then quite obscure Watson and Crick, probably helped give it respectability.

In this connection, it should not go unnoticed that Sir Lawrence Bragg, who less than two years earlier had summarily warned Watson and Crick off DNA, was far from disapproving when they "won" DNA for Cambridge (which, not incidentally, was still smarting from the "loss" of protein structure to Linus Pauling). Bragg even went further and wrote the Foreword to *The Double Helix* — a decided coup for Watson in the face of vociferous disapproval of the manuscript by many of its characters, including Crick who threatened to sue. In this Foreword, Bragg expresses "deep satisfaction" at the "due recognition . . . given to the long, patient investigation by Wilkins at King's College (London) as well as to the brilliant and rapid final solution by Crick and Watson at Cambridge" by the Nobel Committee's decision to split the prize between them; but not a word about Franklin.

So, the standard of morality certainly is low; but I am not sure that this is new. I have recently looked into the origins of my own field of visual pigment biochemistry, which dates back a hundred years, and find there a situation that also is very bad. Being the nineteenth century, of course, it involves only men. In this instance, a discovery by an apparently very modest, young scientist immediately turned into a race with a much older, more established and powerful one. The young man shortly was dead from tuberculosis; the older one continued with the work and entirely overshadowed him.[13]

A final point: At the masthead of his concluding chapter in *The Path to the Double Helix,* Olby has two quotations, one from Crick, the other from Michael Polanyi, a philosopher of science who began his career as a physical chemist. Crick's quotation begins: "The ultimate aim of the modern movement in biology is in fact to explain *all* biology in terms of physics and chemistry." And the rest of it expands on this reductionist paradigm and ends: "It is the realization that our knowledge on the atomic level is secure which has led to the great influx of physicists and chemists into biology."[14] Placed right up against this, Polanyi's quote begins: "The universal topography of atomic particles . . . which, according to Laplace, offers us a universal knowledge of all things is seen to contain hardly any knowledge that is of interest." It ends by extending this recognition to biology: "But now the analysis of the hierarchy of living things shows that to reduce this hierarchy to ultimate particulars is to wipe out our very sight of it. Such analysis proves this ideal to be false and destructive."[15]

Such is the parting of the roads that we confront in biology. The reductionists call the synthesists pessimistic; the holists call the reductionists blind, and even destructive and dangerous. Nothing illustrates the dichotomy more clearly than the current controversy over recombinant DNA. For this technology is a practical application of the reductionist view, that sees nothing improper in regarding living organisms, such as the lowly colon bacillus, *E. coli*, as convenient chemical tools; and that sees no danger in teaching these bacteria in the wink of an eye a molecular vocabulary that other, "higher" organisms have taken eons to learn — and eons during which there has been an opportunity for many trials and many fatal errors. This is the side on which Watson, Crick and most of the other "molecular biologists" line up. On the other side are people who see the organic world as much more than the sum of its parts; who believe that connections and process cannot be understood by isolating things or events; who distrust oversimplifications, and in Chargaff's words, hate "to see *E. coli* impersonating nature. The difference in talents is really too great."[16]

It is not pessimistic to recognize that in the real world everything is connected with everything else; that "isolated variables" are figments of a miseducated imagination. "To light a candle is to cast a shadow," writes Ursula LeGuin in *A Wizard of Earthsea*. This does not mean that we should abandon all attempts at analysis and at science. But it means that we must keep our perspective on what they can tell us. We cannot expect the science of atoms and molecules to unveil what life is.

---

*Following Professor Hubbard's talk, there was a wide-ranging discussion that lasted almost an hour. Despite its general interest, space limitations make it impossible to publish the transcript here. (The transcript has been deposited in the Wrinch Collection.) However, the following exchange is of such historical and scientific interest that we decided to include it.*

*David Harker:* I think that perhaps I should say something, because I know almost all these people you have been talking about. Francis Crick worked in our laboratory for about two years in the '50's in the middle of everything that was going on in England. Sir Lawrence

Bragg found him so talkative that he couldn't think. He had heard that Harker had some money to spend on structural studies of biological materials, and sent him to me. Also, during the '50's, Rosalind Franklin visited us at Brooklyn Poly, stayed in our apartment, and she appeared to me and my wife to be neither uppity nor combative but merely a quiet, modest, and very intellectual scientist involved solely in her interest in her work, and getting support for it. It may be appropriate to say that I think she was a magnificent experimenter, and did produce the best x-ray diffraction patterns of DNA and like material that I had ever seen. (There may be better ones now.) The fact that she did this does not mean that she would surely have worked out the structures. On the other hand, Crick, not a good experimenter, was an extremely skilled intellectual in regard to structures, knowing all the history, numbers, and facts of structural chemistry. He also knew the ways molecules act on each other, understood directed forces like hydrogen bridges, and things of that sort. Might this knowledge (together with the worst data, or no data at all) have produced the structure which he would then set about finding somebody's data to check? . . . Crick, very imaginative, brilliant, a real genius intellectually, and knowing structural chemistry very well, might have produced this structure for DNA from its chemical formula and then would have set about looking for experimental data to check it, and would probably have accepted anybody's that he could have found. (He said things like this.) He also once said, "When the experimental data don't agree with my theories, I think they should be repeated!"

And the real tragedy in this affair is the very shady behavior by a number of people, as well as a number of unfortunate accidents, which resulted in the transfer of information in an irregular way. I think this occurrence is almost unique in my experience of scientists. I think I'm a pretty good scientist. I would never have consciously become involved in anything like this behavior, especially the transfer of information through a privileged manuscript. And I think these people are — to the extent that they did these things — outside scientific morals, as I know them.

*Ruth Hubbard:* I don't really want to get into the question of would she, would he, who did. I think the important thing is that they didn't acknowledge that they knew things that *she* had discovered. The question on the passage of information, I was just thinking about yesterday, and I suppose the only thing one can say about both what Wilkins did and what Perutz did was that they might have taken seriously the fact that Crick and Watson were not supposed to be working on DNA and were working on different things. In fact, Perutz has said that. I can see that Perutz, knowing that here are two people fascinated with DNA, though they are not working on it, when he gets this report, would say, "Hey, look, I just got some really interesting stuff about DNA." Now you know, it is not really the best mores to do that, but I don't think it is totally outrageous, granting that they were interested in it, but it wasn't "their thing" anyway. It seems to me that what was improper was for *them* promptly to *make* it "their thing".

*Carolyn Cohen:* First of all, I think one can characterize this behavior as an extraordinary lack of generosity. I think that this is a general problem in science, but the DNA story is an unusually extreme case. They should have acknowledged where the information for the solution came from and they did not. In fact, the way the x-ray data were appropriated was improper.

And so, I really can't agree with Dr. Harker's point that Francis Crick, because of his great structural knowledge, would have solved it. Even Linus Pauling could not solve it without the necessary x-ray data.

Yet, it is not a matter of *whether* it would have been solved, but really of *when* it would have been solved. This fits into something that is very important — why that x-ray diagram could have been solved in three weeks. It was not the diagram of a crystal. It was a fiber diagram; that is, the x-ray diffraction pattern of a helix. It was Rosalind Franklin's pattern of the B-form (the wet form) that showed the helix most perfectly. This was the wonderful diagram which enabled the basic parameters of the DNA helix to be established — but not the exact structure. The knowledge that it was a helix sprang from the appearance of the pattern, and indeed Crick made major contributions to the analysis of helic differential diagrams. He was an expert in deciphering such patterns. Watson's contribution was the model-building, using the chemical data of Chargaff. He recognized that a specific matching of the purines and pyrimidines allowed a regular stacking of bases at the center of the helix. This two-chain model gave the solution to the diffraction pattern. Now it is perfectly true that Franklin was approaching the problem from a different point of view. She was trying to do it with Patterson models; she was trying to solve the pattern objectively, without model building. In this, perhaps, her instinct was wrong.

*David Harker:* You've made some very important points and I guess I must say a couple of things about them. The kind of diffraction pattern even the most excellent fiber will give (the best I've ever seen is perhaps the one of DNA, and there is a good one of collagen too) cannot be solved in the sense that a real crystal, with three-dimensional periodicity, can be solved if you have a very good diffraction pattern of the crystal. Instead you can look at such a fiber pattern and say that there is a large class of structures which cannot give this pattern, and there is another large class of structures which must be checked against it because they might give this pattern. For every pattern that is compatible with the helix, there are also structures with which it is compatible which are not helical. To say a pattern, a fiber pattern, *proves* that a helix is there (as Crick did in a seminar with us), I object to strongly and I can work out structures, hypothetical structures which would give exactly that pattern, which are *not* helical. So that means you cannot work out the helix from the pattern. You have to *think* the helix, compare it with the pattern, and if it checks throughout with the helix...

*Carolyn Cohen:* This is a technical point. Most polymers do form helical structures but the x-ray pattern alone is certainly not sufficient for a solution. There are a number of helices compatible with the pattern and indeed there are many helical ambiguities. The crucial point was accepting the hypothesis of the helix, and building a model that was *chemically* sound. The chemical part was equally crucial and there the Chargaff data were essential for the base-pairing scheme of the two complementary chains. Model-building is a way of arriving at a good structure, but it gives you a solution that must be checked by other means. Indeed, it was an elegant model for DNA, and one that was later proved using detailed data by more orthodox crystallographic methods.

# LIFE IN THE UNIVERSE*

## George Wald

*It is a great honor and pleasure to take part in this meeting in memory of Dorothy Wrinch. She was a most remarkable woman, our neighbor in Woods Hole on Cape Cod for many years. Both Ruth (Hubbard) and I visited her whenever we could and always enjoyed enormously her keenness, her humanity, her wit. She was an altogether wonderful person. What I'm about to say is a small dedication to Dorothy Wrinch.*

There probably never was a time when it was as important as now for human beings to try to evaluate their condition, to try to understand the nature of our universe, the plece of life in it, the place of humankind in life. We know now that we live in a historical universe, one in which not only living organisms but stars and galaxies are born, come to maturity, grow old, and die. It is apparently a universe permeated with life, in which given enough time, life arises whever the conditions exist that make it possible.

How many such places are there? It looks as though the already observed universe contains at least something of the order of a billion billion places that could support life — $10^{18}$.

Our universe is made of four kinds of elementary particles: protons, neutrons, electrons, and photons. The simplest way of thinking of starting such a universe as this is to begin with equal numbers of protons and electrons which, if they bonded together, would constitute hydrogen. If not the universe as a whole, surely large parts of it began as a mass of hydrogen gas, filling large volumes of space. Over the ages, here and there, quite by chance, a little eddy forms in the hydrogen, which if big enough begins to draw in the hydrogen about it, through the ordinary forces of gravitation, and thus begins to grow. The growing, collapsing mass of hydrogen

*We present here excerpts from the transcript of Professor Wald's talk, prepared by the editor.

heats up, and when the temperature in the deep interior reaches 5 million degrees, the hydrogen nuclei, the protons, begin to fuse to form helium nuclei. But in this transaction a tiny bit of mass is converted to radiation. That radiation, poured out in the interior of what had been a collapsing mass of hydrogen, backs up the further collapse and produces an uneasy steady state. What I have just described, of course, is the birth of a star.

The star, we say, is now mature. It is on the Main Sequence: it lives by the process of fusing hydrogen to helium. The time must come in the life of every main sequence star when it begins to run short of hydrogen, and with that it begins to produce somewhat less radiation and to collapse again, and to heat up some more. When the temperature in the deep interior reaches 100 million degrees, the helium nuclei begin to fuse. And that is how carbon, oxygen, and nitrogen come into this universe.

These processes result in a new outpouring of energy that not only backs up the further collapse of this star but puffs it up to enormous size. It becomes a red giant, a dying star. Red giants are in a delicate condition — they are always distilling off some of their matter to become part of the enormous masses of gases and dust that fill all interstellar space.

And over the ages, quite by chance, here and there an eddy forms in the gases and dust and a new star is born. But these later generation stars, unlike the first generation, have carbon and nitrogen and oxygen. We know that our sun is such a later generation star, because we are here.

There is nothing as much like a living organism as a star. Like a living organism, a star has its metabolism. We live by burning hydrogen, they live by fusing hydrogen. Like a living organism, a star is born, matures, grows old, and dies. Like a living organism, a star has pathologies. And the stars greatly resemble living organisms in their composition. They and we are made of the lightest elements. The planets are altogether different. They are the dead ashes of the universe, made of iron, nickel, silicon, aluminum. They are dead, but we and the stars are alive.

The stars are much too hot to have the atomic nuclei in them gather electrons about themselves in orderly ways. For that to happen, one has to go to the cooler places in the universe, the planets. There those electrons can interact with one another to form molecules. When our planet Earth reached its present form about 4.7 billion years ago, molecules began to form, including those special ones made of hydrogen, carbon, nitrogen and oxygen that we call organic because they have such an intimate association with life. Organic molecules kept forming over the ages so that the oceans became an increasingly concentrated and complicated organic soup. And — going very fast — somewhere, sometime, or perhaps several times in several places, an aggregate of organic molecules in sea water reached a state that an experienced biologist, if he had been around, would have called alive. This happened on this planet something like 3 billion years ago; and having acquired a foothold, life began to flourish on the earth.

Life is a wonderful thing whenever and wherever it appears in the universe. Life seems to represent the highest state of organization that matter achieves. It has a special characteristic that is very strange: individuality. Not only is every living organism different from every other, but it differs from itself from moment to moment. (That characteristic we share again with the stars.)

Something like 2 million years ago, human-like creatures appeared on the earth. About 400 years ago, a collection of molecules, organized as William Shakespeare, wrote *Hamlet*. I don't say that kind of thing to disparage human beings, but to exalt the molecule! When you see a collection of molecules writing *Hamlet,* you see what molecules, properly organized, can do! You have learned a little special biochemistry.

These wonderful things happen as the result of the operations of a principle of design that is enormously important to all of us. It was described a little over 100 years ago by Charles Darwin and called by him Natural Selection. I like to call it organic design. In an increasingly technological world, one quite easily makes the mistake of thinking that all the design we see about us is technological design. But in fact, living creatures were made through an entirely different process of design, amost the reverse of technological design. Technological design begins with specifications. One lists the specifications and then does one's best to achieve them. But in natural selection, in organic design, there are no specifications.

Organic design has three components. First of all, a ceaseless outpouring of variations, advantageous and disadvantageous. Second, some mechanism of inheritance (biological, social, cultural). And finally the selective component: a principle of competion. To oversimplify, any population of living things tends to fill and then to overfill its environment, and then to compete for the needs of life. Those organisms and those characteristics of organisms that *work a little better* survive — and those that work a little less well decay and disappear.

The whole process is one of editing, rather than of authorship, and people who like such phrases could speak far better of the Great Editor of our being, rather than its Great Author.

Let me interpolate a word at this point. Ruth and I are among the few scientists who speak up in opposition to the "great breakthrough" that has recently occured in biology, called recombinant DNA. My own principal interest in opposition to the use of this technology does not involve its safety but something that hardly has come up for discussion at all. What recombinant DNA, the gene-splicing technology, really means is the technologization of life. It is the re-designing of living organisms to specification. That, I think, is the really profound danger in the pursuit of this technology.

Now I come to my main theme, which is that if one were to change any one of a number of properties of the universe, the life which now arises inevitably would become impossible here or anywhere. I want to explain this by climbing up the scale of organization of matter.

I will begin with the elementary particles. You know, the strangest experience that a scientist can have is to recognize that something one has always been taught, and has taken for granted, represents a question. The question I am about to introduce to you struck me after lecturing on it for 13 years! It is: How does it come about that two elementary particles as completely unlike each other in every other regard as a proton and an electron have the same numerical charge? In 1959 two of the most distinguished astrophysicists alive published a paper in which they proposed that in fact the proton and the electron differ in charge by an almost infinitesimal amount. That would have explained quantitatively the expansion of the universe: everything would be charged in the same sense, and since like charges repel, all matter would repel all other matter and the universe would expand, — as indeed it does. The only trouble is, it wouldn't do anything else. That tiny difference in charge would be enough to overwhelm the forces of gravitation that bring matter together. And so we would be in an expanding universe, but with no planets, no stars, no galaxies, and worst of all no physicists.

Let us go up a step to atoms, elements. There are 92 natural elements but 99% of living material is made of just four of those — hydrogen, carbon, nitrogen, and oxygen. The study of chemistry is based on the Periodic System of the elements: we write the elements in horizontal rows and one is taught that if you go vertically down the columns the elements repeat their periods. But the truth of the matter is that the elements of the first two periods — which include hydrogen in the first period, and carbon, nitrogen and oxygen in the second period — have absolutely unique properties, not repeated again by any other elements. For example, anybody who has ever studied organic chemistry knows that there are not only single chemical bonds, but also double bonds and triple bonds. But only carbon, nitrogen, and oxygen form anything other than single bonds. Consider the molecules carbon dioxide and silicon dioxide. In carbon dioxide, the carbon is attached to the two oxygen atoms by double bonds, and this saturates its combining capacities. It forms a gas that dissolves in the waters of the earth, and from this living organisms get their carbon. But in silicon dioxide, the silicon is attached to the two oxygens by single bonds; its bonding capacity is unsatisfied. So each silicon dioxide latches on to a neighboring one, and that to the next one, and so on, and when you have finished you have quartz.

Let's go up a step to molecules. By far the most important molecule in life is water; and water is the strangest molecule in the whole of chemistry. There is nothing else like it anywhere, and its strangest property is that ice floats. Suppose it didn't. Then freezing would start at the botton, not at the top, and bodies of water would freeze solidly. If this happened on any planet in the universe, and even once in a billion years there was a cold spell that froze the water, that would be the end of life on that planet.

Let's go up another step. What does it take in a planet to support life? Well, it must be about the right size, about the right temperature, and receive about the right amount and quality of radiation from its star. Why the right size? Because the only thing that holds an atmosphere to a planet is gravitation. Our planet is the right size; Mars is too small. Jupiter is too big. Most of Jupiter is a solid atmosphere, and life is hard in a solid stmosphere. How about temperature? Together with other considerations, we need a temperature at which water will stay liquid. As for the radiation, that is somewhat easier to arrange. Thus in any solar system in the universe, the overwhelming likelihood is that there will be just one planet with properties that can support life.

Let's go up a step. What does it take in a star to have a planet that will support life? A number of things. It has to be a main sequence star, one that lives by fusing hydrogen to helium. It has to be a single star, not a multiple star. (Planets make very irregular orbits around multiple stars, so they go through enormous swings in temperature that almost surely would be incompatible with life.) Then the star should be the right size, the right mass. The bigger the mass of the star, the higher the temperature at which it settles down on the main sequence, and hence the faster it uses up its hydrogen. A star twice the mass of the sun would stay on the main sequence only about a billion years, but it took something like a billion and a half years to get life started in this solar system.

And finally, there is a cosmic condition. If the universe were not expanding, the light we receive from the stars would be much greater. In fact the sky would be ablaze with light night and day, and every planet in the universe would be heated up to incandescence, because it would absorb that light until it was radiating light as fast as it was absorbing it. And life is hard on an incandescent planet.

Let me summarize this story now. It is telling us some rather strange things. It's telling us that we are living in a very special kind of universe. It takes no great imagination or skill to dream up other universes which would be perfectly good universes, but with no life. If the electron and proton didn't have exactly the same numerical charge; if hydrogen, carbon, nitrogen and oxygen didn't have the unique properties they do; if ice didn't float; if this were not an expanding universe; it might be a perfectly good universe, but without life.

Human beings and their like are a great event wherever and whenever they occur in this universe. Human beings have demonstrated, on this planet, that an animal can produce art and science and technology. But we have reached the astonishing point of not only dominating life on this planet, but threatening to extinguish ourselves. There is no assurance I can find for myself or for you or for my children that we will be in physical existence 5, 10, or 15 years from now. A half hour's interchange between United States and the Soviet Union of the present stockpiles of nuclear weapons would be the end of the human race.

We are not made to be exposed to stuff like that. We, you and I — the matter that makes our bodies was cooked in the deep interiors of former generations of stars that have died, garnered from all the furthest reaches and corners of the universe over endless spaces of time. 20 billion years of the universe, almost 5 billion years of the earth, 3 billion years of life, 2 million years of human creatures, 10,000 years of civilization. And 32 years ago the first atom bombs were dropped on Hiroshima and Nagasaki, killing some 250,000 people. By present standards, those two bombs were pitifully small. Today they would be rated as tactical nuclear weapons.

And that is now our problem.

# II. The Dorothy Wrinch Collection

# THE DOROTHY WRINCH COLLECTION

The Dorothy Wrinch Collection includes over twenty large boxes of correspondence, papers, notes and miscellaneous items, in addition to many notebooks and annotated reprints of articles by herself and other eminent scientists. The bulk of the Collection consists of letters written to and by Dorothy Wrinch. There is correspondence with G.H. Hardy, Irving Langmuir, Eric Neville, Michael Polanyi, Harry Sobotka, Dorothy Hodgkin, Martin Buerger, Isidore Fankuchen, J.D. Bernal, David Harker, Niels Bohr, J.D. H. Donnay, and many more. Some of these letters are purely scientific, while others are both scientific and personal.

The cyclol controversy of the late 1930's is well documented. Dorothy Wrinch's subsequent joint appointment to Smith, Amherst and Mount Holyoke Colleges is fully described, documenting the first efforts toward cooperation in the Valley. Her Smith class notes, lectures, and examinations are included, as well as her ideas for new courses and her plans for the development of a department of molecular biology. The Collection contains grant applications and progress reports to foundations, and detailed communications with many colleagues. There is an unpublished manuscript criticizing the Watson-Crick model for the structure of DNA. Letters and notes relating to her three monographs are also preserved here.

There are many notes which she wrote to herself, ranging from scientific to personal philosophies, from spontaneous thoughts jotted down on scraps of paper to well-organized theories. They record a struggle to unite specialized scientific disciplines and a battle against scientific isolation.

*Claire Sullivan '78.*

# THE WRINCH PAPERS

*Mary-Elizabeth Murdock:* First of all, I should like to give you a general idea of what the Sophia Smith Collection is, and its significance. Second, I want to comment on some of the problems that we run into as we process papers. Finally, we shall talk about the Dorothy Wrinch papers, their importance, and various things that Claire has found and you have seen, enlarged for the display for this symposium. [See "Selections From The Wrinch Papers," pp. 180-189. Ed.]

So let me start right away with a very small talk about the Sophia Smith Collection. Established in 1942, it was named in honor of the founder of Smith College. From the very beginning, its purpose has been to document the lives of distinguished as well as "everyday" women with the view toward correcting the impression that women have not really accomplished much throughout history. Women, in fact, have contributed many skills in a variety of ways. We have, for example, some papers of women in science, politics, fine arts and humanities. Probably our strongest very large collection is the Garrison Family Papers that document the involvements of a major American family in antislavery, women's rights and other reform movements in the 19th into the 20th century.

*Martin Buerger:* Do you limit yourself to Smith?

*Mary-Elizabeth Murdock:* Absolutely not. I am glad that you raised that point. The Sophia Smith Collection does contain the papers of some Smith alumnae, of course, because Smith has a habit of turning out many distinguished people, but we do not seek papers solely with alumnae emphasis. We screen all proposed collections quite carefully, whether they are alumnae papers or not. The Schlesinger Library at Radcliffe, by the way, is the other comprehensive women's history archive comparable to this one. The main difference between the two archives is that the Smith collection does have an international emphasis. For example, we have fine English suffrage materials that are fascinating from a research point of view because they reveal in no uncertain terms the subtle differences between the American and the English suffrage movements.

Well, to get back to the Sophia Smith Collection. Perhaps you might wonder why we bother to collect women's papers. Let me offer a few of my thoughts on that subject. I think what we do when we collect the papers of women is to reveal women's role in the life of a particular period of time. From reading the Garrison Papers, for example, one may get a sense of what it was like to be a woman in the 19th century — the conflicts between staying in the home and wanting to attend antislavery meetings — or wondering why slaves were becoming free and women weren't — that kind of thing. Fascinating portraits of the everyday lives of women emerge from the papers we collect. In a real sense, papers can almost bring people back to life. We may understand how they lived, why they made the decisions they did, the nature of their work, the relationships they had with other people, what was important to them — and what mattered to other people about them. For all of these reasons we try to collect women's papers. In doing so, we are also attempting to revive a segment of history that has not received adequate treatment, and that is the role of women in the United States and throughout the world. Feminists often claim that women have been totally forgotten. I don't really think so. Our papers in the Sophia Smith Collection indicate that women have not been forgotten. The problem lies, perhaps, in our formal educations that all too often fail to reveal the fact that women have, indeed, achieved distinction in many fields. Preserving papers will help to correct this serious oversight in the future.

I think we ought to turn now to the Dorothy Wrinch Papers, which we were extremely happy to obtain. There was some question, while Dorothy Wrinch was still living, whether the Sophia Smith Collection would get them. Often when someone has a valuable collection of papers, they tend to be terribly possessive of them. They ponder questions such as, "Where should I place these things? Where would they be the most used?" I shall be very candid and say that when the question of placing the Wrinch Papers in the Sophia Smith Collection first surfaced, I had my doubts about it. I wondered if the complexity of the subject matter would be too advanced for an undergraduate institution. Colleagues managed to persuade me that I was wrong, so I have changed my mind. Now I think Smith *is* the place for the Dorothy Wrinch Papers. This college offers a superb education in the sciences, and I know that Mrs. Wrinch was interested in working with the students here. She had a way of collecting protegees around her. For these reasons, the Sophia Smith Collection does appear to be the proper repository for her life work. It would have been impossible, however, for anyone on my staff to cope with the highly technical, sophisticated nature of the material. But Mrs. Senechal was very helpful in counseling us and recommended our able friend Claire Sullivan here, who has organized this collection enough so that we can now transfer it to our quarters, shelve and catalogue it according to our standard procedures. I think Claire wishes to comment on some of the remarkable things that have emerged in the Dorothy Wrinch Papers.

*Claire Sullivan:* As Miss Murdock said before, the Wrinch Papers do indeed show a great deal about Dorothy Wrinch and what she was like as a person, and also her style of scientific study. Included in the collection are many personal notes and jottings that she wrote to herself which I find are very revealing about Dorothy.

*Martin Buerger:* Why did she file such things away?

*Claire Sullivan:* I don't know why, but I can tell you how: in the most crazy manner! She had them all over the place. Some of them are on little pieces of paper. This one fortunately has a date on it, so I have some clue as to where to put it. But a lot of them are on little scraps of paper, stuffed in envelopes or inside a reprint of somebody's — all over the place. I try to pull them out and put them in special categories.

*Mary-Elizabeth Murdock:* I might just add a comment here. We find that very often the notable women whose papers we process have a common characteristic — a sense of history. They believe that they are women of significance and that they've made a contribution to society. There is rather clear evidence that Dorothy Wrinch felt this way about herself. Probably she made a conscious decision not to throw anything away.

*Claire Sullivan:* Yes, they are very interesting. A lot of them reveal a very logical method of thinking, even when she was just playing with thoughts, not scientific at all. For example, this little piece of paper says, "A makes his living out of B (among others). A calls on B in his course of making his living out of him, thereby preventing B from making his living, without which B can't be the material of which A or anyone else makes his living." This is typical of a mathematical sequence of thought, even though she apparently was not dealing with a mathematical or scientific idea, here.

*Mary-Elizabeth Murdock:* It is a good thing we don't know who A and B are!

*Claire Sullivan:* Yes, certainly. Although there are often little initials when she talks about people, you usually have no idea who they are referring to.

Dorothy Wrinch was also a mother, she had a daughter Pamela, and she raised the child singlehandedly since she was divorced for most of Pamela's childhood. She remarried later, but a lot of time was spent without a helping husband. A lot of notes are on thoughts about relationships between mother and child, and things like that. Here is an example: "the thing in the *New York Times* about the child who transgresses, and the mother does not reprove her, because the child will get a feeling of guilt of having done wrong, so the mother really gets the guilt feeling . . . This is an absurdity. The point they say is: let the child have a feeling of guilt commensurate with the wrong performed. Thus, if she has not cleared up her toys or whatever, let her guilt be small, not the guilt, say, of having committed a murder . . . Do you think the mother should make the child feel guilt when she transgresses or NOT? Answer yes or no. (NB. Compare the yes or no questions and examine in USA or possible yes to one of these three questions.)"

Dorothy Wrinch lived in Hubbard House after her second husband, Mr. Glaser, died, and she had many opportunities to observe Smith life and students. I interpret this note as being some comments of hers on Smith life: "Yes, ah yes indeed, they do build character. As for example: having a phys ed and med department which holds with — insists on — plugging for removing every spare ounce of fat, and then serving them whipped cream approximately 10 times a week." (And incidentally, I can tell you that is true, having spent a year eating in one of the college houses.) "This is a remarkable battlefield atmosphere which should prevail (as it does sometimes, though not always) on each of the 10 to 14 times per semester — which is part of a grand design for building strength of character.

Actually it is due to the help of those who insist on this diet, a collection of employees who regulate all the details of their employment by a union already 1 year old..." It was written in 1959 and it still holds (I think) in 1977.

There are also a few notes that seem to indicate feminist inclinations, or comment on the typical stereotype of a female. You may not get the same impression from this note, but this is what I think. "To you, A B C D is not impossible — We stand ready to serve your every need to make a more beautiful, a more elegant, a more sophisticated, and a more irresistible you . . . To whom, for whom, only a pedestal is an appropriate seat. Look at those girls — yes there are some — look at her hairdo, her clothes, her . . . — and look at her test tubes and oh that apparatus, so confusing looking and lacking any charm. What could be the fascination of such a messy looking bit of stuff?" There is an awful lot of sarcasm but it is one of my favorite notes.

From what I can tell from these papers, Dorothy Wrinch was a very sensitive and perceptive woman, not only with respect to science. She seemed to have some kind of "second sight", which she used to see what people really meant, rather than what they actually said, or wrote. She often wrote comments depicting what she saw with her "second sight". These comments, I think, are very sharp and very witty. For example, I'd like to read to you my favorite piece in the whole collection. This letter was written by Francis O. Schmitt to Professor Andrews about Dorothy Wrinch. Professor Andrews sent a copy to her.

*David Harker:* Andrews of Hopkins?

*Claire Sullivan:* Yes, it was in 1941. "Dear Professor Andrews, I write to thank you for your letter regarding Dr. Dorothy Wrinch. Several appointments have already been made to strengthen our biophysics staff at M.I.T. and it is possible that one or two further appointments may be made, though I doubt that this will occur in the near future. We shall be very interested in the question of protein structure, but chiefly from the experimental, rather than the theoretical approach. I have a very high regard for Dr. Wrinch, and I feel that she has made a very stimulating contribution to the theory of protein structure regardless of the eventual disposition of the cyclol theory. Her originality is most refreshing and I enjoyed chatting with her on biological and chemical matters. I can foresee difficulties, however, in fitting her into any permanent sort of staff position in a department of Biology such as we have at M.I.T. It may be that she will eventually find a solution to her problem by taking a position in a mathematics department under conditions which will permit her to devote considerable time to investigations of protein structure. In any case, if anything presents itself which would seem to be suited to her needs, I shall be happy to get in touch with her. Sincerely yours, Francis O. Schmitt." The following note was written [sic] by Dorothy and attached to the back. "This letter provokes the following comments: His interest in protein structure chiefly from the experimental rather than the theoretic approach shows a complete ignorance of the structure of native proteins. This is not his fault. Knowing nothing of structure, and depending for his information on current textbooks, he couldn't expect to understand them. But you might have thought he could read the message in my papers, namely that no molecules with thousands of atoms and well-defined structures can be dealt with except by taking in turn all possible structures, theoretically constructed, and testing them in turn. His researches will continue to throw as little light on fundamental structural problems in biology as in the past, no doubt. He enjoys chatting with me on biological and chemical matters. This is very nice. Why do I chat with such creatures? I should read their works with care in case, by some extraordinary chance, some of their brute facts have structural significance, but I should not give them the benefit of chatting with me. His right to have any opinion on the ultimate disposition of the cyclol theory seems to be limited to his capacity to quote at me remarks said to have been made about it by Bragg and others. He understands nothing about it, see first paragraph. 'Difficulties' means apparently that I don't know enough biology or what??? I suppose I should be out of place in a society with their level of intellectual understanding. 'A solution to *her* problem...' Apparently the question is how to get a living for a woman who after all has a right to live. It never occurs to him that I have a contribution to make or may even have made one, and that it is *their* problem to see that I have the opportunity to do my best work under the best conditions, for the sake of the progress of science. My opinion of F. O. Schmitt is confirmed. He is a facile experimentalist. His forte is making gadgets. God help biology so long as this is the type of person chosen to break new ground..." I think this is fairly typical of her reactions to other people.

*Mary-Elizabeth Murdock:* As I commented earlier, as you process papers, you do get insights into people's minds and how they feel about things, how they relate to other people. I think this is a perfect example.

*David Harker:* It is exactly the atmosphere she created in our little group at Hopkins. It makes people who absorb it into a pretty snobbish school fraternity. And it shows exactly why she didn't get along any better than she did. Not because she was a woman, but because it was only *her* idea that had to be accepted.

*Mary-Elizabeth Murdock:* Do you have anything anywhere in there that demonstrates how she related to women colleagues?

*Claire Sullivan:* She had an off-and-on relationship with Dorothy Hodgkin, from what I can tell. Was it always off?

*Marjorie Senechal:* It's not clear, but it wasn't always on.

*David Harker:* I don't think they were inseparable, by any means.

*Mary-Elizabeth Murdock:* So we have a question here of competition, regardless of sex.

*David Harker:* I thought so. I think the sex theme is something one can discuss, but I think it's been overestimated during these meetings here.

*Mary-Elizabeth Murdock:* So it's essentially a question of competition of people with people.

*Elizabeth Moore:* And of particular personalities. She was very dynamic. She was the bulldozer type, wasn't she? She'd tend to bowl you over with her energy, her ideas. It was hard to get a word in edgewise.

*David Harker:* It sure was. I was thinking this noon, just after the meeting [Ruth Hubbard's lecture — Ed.], Francis Crick and Dorothy Wrinch are very parallel scientific personalities. The same dominance of the conversation and insisting on talking about *their* interests, not *your* interests, and the same combinatorial minds, brains. I think the main difference is that Crick, when one of his models didn't work, was willing to try another one. During this DNA business, he did try several models, and one of them seemed to have all the properties of a good model and he promoted that strongly —

just the way Dorothy Wrinch promoted her cyclol stuff. But he had very good luck, his checked out; hers didn't.

*Mary-Elizabeth Murdock:* I rather think that if one's exploring a new field, the structure of proteins, maybe you have to be somewhat forceful (shall we say?) to get anyone to listen to you. I think we find that in many notable women.

*David Harker:* Pauling is that type too.

*George Fleck:* Crick was able to get a reputation for his model and, later on, data to support it within a reasonable number of years, whereas Dorothy had a model which maybe was right and maybe was wrong, for a very long period of time, with no data that would really push it off center.

*David Harker:* I was intended to be the person who would check her model so I investigated what would be involved in doing the experiments and x-ray diffraction. And I guessed 10 years, with giving up teaching and having complete support, and I would have to have assistants who were not graduate students, as I could not guarantee them Ph. D.'s. And so I found it impossible at that time. In 1950 I did get the opportunity to do this kind of thing and it cost more, by a factor of two, than I expected — although there wasn't much inflation then. And it took 17 years instead of 10 — and that was 10 years later, almost.

*George Fleck:* It is not too hard, psychologically, to hold a hypothesis for a year or two, and then say, "Oh well, another one will do." But when you have a hypothesis which is held in limbo, and experiments won't do anything one way or the other to it, and you have fought hard for it, and the decades go by, it becomes harder and harder to say, "Well, so much for two decades of fighting, I'll try a different model." It's easier if the feedback is earlier.

*Claire Sullivan:* I have something here that supports this. It's a letter to Dr. Clowes, dated 1955. "During last week I finally succeeded in getting hold of a long and very complicated article by Professor Stoll in which he demonstrates the existence of the cyclol groupings in the five natural alkaloids of Ergot which have a peptide portion. . . Undoubtedly the results are of great importance in themselves. Also they interest me very much because I had always expected the cyclol structures would turn up first in work with hetero- and polycyclic compounds. However, I feel in great need, in my isolated position, of an authoritative and unbiased opinion as to the significance of these results in regard to my cyclol theory of protein structure. It has been a very, very long wait — from 1936 up unto now — to have the existence of the cyclol bond confirmed. The total lack of interest among protein chemists in the idea of any multiple bond between amino acid residues and the native proteins, cyclol or other, has seemed to me to be a most unfortunate thing. However, all this long time I have been powerless to alter the situation. I am naturally very anxious now to form a sound judgment as to whether Stoll's results do or do not alter the case regarding the cyclol theory for proteins. If they do, I would then be able — after this very long and frustrating period — to make new plans to attack the chemical aspects of protein structure.... Sincerely yours, Dorothy Wrinch. September 30, 1955."

*David Harker:* Protein chemists were always looking for what happened inside proteins, and they did not start out to neglect the cyclol theory. It just turned out not to be the structure of proteins.

I want to ask Dr. Buerger, whether she didn't indeed start him off on at least the basis for his method of finding crystal structures, the implication theory, which I think, has become the theoretical basis of all modern methods of determining structures directly.

*Martin Buerger:* I believe that Dorothy Wrinch's influence on direct methods is not along the lines that you imply. I believe I first met her at one of the summer meetings at Gibson Island on x-ray and electron diffraction, sometime in the late 1930's or early 1940's. At the time I was already interested in applying Harker sections to the analysis of the crystal structure of the mineral nepheline, $KNa_3Al_4Si_4O_{16}$. As I recall it, Dorothy Wrinch had been scheduled to talk about vector sets but, instead, talked about Fourier transforms. Perhaps it is best to briefly review here the history of direct methods to see what Wrinch's influence was.

In 1929 Bragg and West discussed the representation of the periodic structures of crystals by Fourier series whose coefficients were the diffraction amplitudes. But it quickly became clear that such solutions were not generally possible because the coefficients are, in general, complex; their magnitudes can be measured from x-ray diffraction records but their phases cannot. This situation led A. L. Patterson to investigate what *could* be learned from the amplitudes alone. In 1934 he showed that a Fourier series whose coefficients are the $[F_{hkl}]^2$'s reveal the vectors between all pairs of atoms in the crystal structure. Then, as early as 1936 David Harker pointed out that certain sections of the three-dimensional Patterson function normal to a symmetry axis ideally contain maxima whose vectors from the origin are those between atoms related by the symmetry of that axis. As I began to apply Harker's discovery to the solution of the structure of nepheline, it became obvious that, if the map of the Harker section were reduced in scale and rotated by amounts which are simple functions of the period of the axis, the result ideally is a collection of peaks whose locations mark out the positions of the atoms of the structure as projected along that axis. This relation holds in its simplicity for some purely axial space groups; in others there are complications due to ambiguities of 2, 3, 4 and 6, and if there are operations of the second sort, there are, in addition, reflection "satellites". The implication function was described in J. Appl. Phys. 17 (July, 1946). The publication incidentally pointed out that the space groups of almost all symmetries can be determined from the Patterson function by making use of the satellites. About three years later D. Rogers and A.J. C. Wilson showed that symmetry elements could also be determined by statistical methods corresponding to Patterson functions lacking geometry.

By December 7, 1941 not only Europe but also the United States were occupied by World War II. For the duration, we all found ourselves with specific war duties; mine was to teach physics instead of mineralogy and petrology. But I did manage to find a little time to continue working on the structure of nepheline so that, when the war ended, I presented the analysis of the structure at the Lake George meeting of the ASXRED in June, 1946. The use of the implication function for this solution and its obvious application to the solution of other symmetrical structures caused considerable discussion. In particular Harker's discussion recognized that the success of the method implied that the phases of the amplitudes were, somehow, contained in the collection of their magnitudes. By the time that the ASXRED met the next summer in Ste. Marguerite, Canada, Harker and Kasper presented their inequalities, and shortly after that they solved their first crystal structure by using these inequalities. In subsequent years, Jerome Karle and Herbert

Hauptman studied the statistical relations between magnitudes and phases of x-ray diffraction amplitudes, as did many other investigators. Eventually all these developments led to the symbolic-addition method which has so successfully solved many crystal structures.

A second aspect of the solution of crystal structures came shortly after the Lake George meeting. I received an invitation to give a series of lectures on vector methods at the Faculty of Sciences of the University of Rio de Janeiro. I was more or less acquainted with Dorothy Wrinch's 1938 and 1939 papers in which she discussed analytically some aspects of the Patterson function.

In my lectures I was able to show that, by making use of the idea of the appearance of one point of a crystal structure as seen from another as a kind of product, which I called an "image", it was possible to develop a simple algebra for dealing with the points of a Patterson diagram. By using this algebra to represent the geometry of these images, there were several ways of solving the Patterson map for the locations of the atoms making up the crystal structure. On returning to the United States I sent the manuscript of "Vector Sets" to Acta Crystallographica in May, 1949, but the editors managed to hold up its publication until March, 1950, when it appeared in the same issue with short notes calling attention to less general solutions by two European crystallographers. The next year, however, Acta Crystallographica published my paper "A New Approach to Crystal-Structure Analysis" in which the actual observed Patterson function of a crystal was transformed into an approximation to the electron-density function with the aid of some new devices such as the minimum function. This paper, therefore, established a route through Patterson space rather than through reciprocal space for direct methods.

Returning to Dorothy Wrinch's contribution to direct methods, I believe her third paper (Phil. Mag. 27, January, 1939), provided geometrical keys to the relation between a set of points and its Patterson set. In that paper she did not claim to present a specific solution to the Patterson function for she states that "the methods here introduced allow any vector point diagrams to be analysed into a finite number of alternative point sets (rarely more than a very few), one among which must represent the structure of the crystal . . ." [p. 99]. Her keys provided me with guides in devising several readily understood and easily illustrated routines for finding the original set of points from its Patterson set by methods which exhaust the Patterson set.

*Mary-Elizabeth Murdock:* Claire, would you care to comment on some of those grant applications and rejections that might be interesting?

*Martin Buerger:* May I first comment regarding the little notes which Dorothy Wrinch wrote and kept? This practice may have been a way of satisfying her desire to respond to a criticism, and yet not sending an irritating letter to her critic. I know that I've done that lots of times! Once I wrote a preface that was a scathing collection of remarks. But I had the sense to throw it in the wastebasket; nobody is ever going to find that on me! But Dorothy saved every scrap. I really think that those notes are interesting in the sense that this is what she would have written if she didn't mind insulting the person.

*Mary-Elizabeth Murdock:* What do you suggest restrained her from just telling them?

*Marjorie Senechal:* There is one letter, to J. D. Bernal, which is unbelievable in its scathing criticism. One would think, because of the way it is typed, that it didn't get sent. But then there are later letters to other people in which she said, "He never answered my letter, I don't understand why!"

*Claire Sullivan:* She received quite a few harsh comments from people. I have a few here from referees for publications. This is the *Journal of Chemical Physics*. One reviewer said, "The views of Dr. Wrinch on protein structure have not been accepted by any responsible worker in the field. There is no significant experimental evidence for cage structure, and the claim that the structure proposed by Sanger for insulin can only exist after degradation is too extravagant to be accepted without very strong evidence in its favor. This paper supplies none of this and is so vague as to be worthless." And the other one says: "Cyclol structures have been described in the literature and discussed for twenty years. [This was in 1957.] No significant information supporting these structures has been found. There is strong evidence that proteins contain polypeptide chains as polypeptide chains. The present manuscript, in my opinion, presents nothing significant. I recommend that the article not be accepted for publication."

*David Harker:* That sounds like Pauling!

*Claire Sullivan:* Here is one from Pauling! This is a letter to William Scott from Pauling in 1956, and he had apparently inquired of Dr. Pauling his opinion on some certain aspect of Dorothy's work. It says: "I do not know about recent developments of the Wrinch protein theory. The papers that Miss Wrinch has published during the last few years and that I have read, have not been convincing. I remember in particular a paper in which she presented an argument to show that the alpha helix could not be present in some globular proteins — hemoglobin, I think. I remember that there was a serious flaw in her argument." He didn't go into too much detail, because it was to a colleague of hers.

*Martin Buerger:* Was this paper suppressed as a consequence of these?

*Claire Sullivan:* I am not sure. The manuscript was entitled "Towards the placing of amino acid residues on the cage model for insulin." I don't think it was published.

*Elizabeth Moore:* She must have had the energy of an ox. Think of just the energy it would take to do all those papers — never mind the brilliance of her mind.

*Martin Buerger:* There were six and a quarter per year for 47 years. That's terrific.

*David Harker:* It's interesting that in the '50's both Pauling and Bragg hypothesized structures for hemoglobin with parallel $\alpha$-helical rods, and everybody was taking them very seriously. That was just about as wrong as Wrinch was. But of course when the data indicated the true structure, they forgot about this. You know, I don't think either of them ever retracted.

*Claire Sullivan:* Also well documented in the collection is the vicious argument between Linus Pauling and Dorothy Wrinch that was taking place in the late 1930's. There is a memorandum entitled "Notes on the talks between Dr. Dorothy Wrinch and Linus Pauling", which took place January 26 and 27, 1938, at Cornell University. This is the discussion that she came over specifically for from England, and apparently it wasn't too friendly.

*Margaret Glennon:* When I came here in 1940, Pauling had just been here as a visitor and then Dr. Wrinch came in 1941. We heard a lot about Pauling from Dr. Wrinch.

*Marjorie Senechal:* The letter you started to read from Pauling to Scott describes a little bit about that meeting. The way he described that meeting in 1938 was very interesting to me because it seems to indicate that there was a real clash of attitudes between the chemists and the mathematicians.

*Claire Sullivan:* Yes, it does. Continuing with the same letter: "A number of years ago I made a thorough study of Mrs. Wrinch's ideas." [He always refers to her as Mrs. Wrinch, not Dr. Wrinch.] "The Rockefeller Foundation had asked her to come to talk with me, and I made a report to the Rockefeller Foundation. The conclusion that I reached then was that the ideas that she had at that time were in considerable part self-contradictory. When I discussed her theories with her, especially the self-contradictions and the contradictions with experiments, she abandoned item after item until at the end there was almost nothing left. The contrast between the extensive claims that she had made in her papers and the very small amount that she was willing to defend caused me to decide not to make an effort to find the reliable parts of her later publications. There is, as far as I know, nothing wrong with her mathematical work on Fourier transforms."

*Marjorie Senechal:* The thing that interests me so much about that comment is that a mathematician who proposes a structure, as she did, would never dream of saying "This is what actually fits nature." He would say, "Here is a possibility. If you start now to work out the details, you will modify it here and there, but this is a way to look at the problem." If a person said, "This doesn't work and that detail doesn't fit," the mathematician would reply, "Well, we'll fix that." This could be interpreted by a chemist as abandoning positions and so forth, and being left with nothing. That could very well have been part of her trouble: coming up against a wholly different way of looking at models and scientific ideas.

I have read Dorothy's early philosophical papers. She was a philosopher and mathematician before she turned to chemistry and biochemistry. I knew she had written on the scientific method, and was curious to see what she thought about it. I didn't read all of them, but I did find one, written in 1928, that was very interesting. In it she said that the purpose of theoretical science is to find all the alternatives that could possibly explain a set of facts, and a scientist's job was to go through them and find out which ones work. She really did seem to believe this. She used many examples from physics, where that kind of thing is perhaps a little bit more feasible. This was her attitude toward science, and in her early papers on the cyclol theory, she said that this was what she was doing. She said, essentially, "There are many possibilities and here is an example of one." This is the kind of thing that a mathematician would do; a mathematician would say, "I know that there are many such structures, here is one that I can produce for you, there may be others, but this is an example of what I am talking about." The people who then got enthusiastic about her work didn't notice this. Lindemann, for example, wrote in support of her two foundations, saying, "She's proved that there is only one way to do this." But, she hadn't proved that and she hadn't said she'd done that. But — this is my own reading of her papers — as the controversy grew and she defended herself, she appeared more and more to come to believe in what she was doing, in a way she hadn't before. By about 1940, she was saying things like *"I'm right!"* Before that, she didn't talk that way, and I wonder if it wasn't the experience of having to defend herself to the world, or putting herself in that position, fighting on every conceivable little detail of the model, that somehow got her locked into thinking that she was right and that everybody else was wrong.

*Evans Hayward:* Don't you think that she had a terrible chip on her shoulder?

*David Harker:* In 1939 and '40 she had a chip on her shoulder. All the time!

*Marjorie Senechal:* No, I don't mean that she didn't have a chip then, but it seems that it did get worse.

*Evans Hayward:* She didn't have what I would call an open mind when I knew her, which must have begun in 1941. This doesn't mean that I didn't think that many things she said were valuable, but she was quite biased.

*Elizabeth Moore:* Could I make a comment? I am one who knew Dorothy and really cared about her; she was very sweet to me as a student and very nice. I actually knew nothing about the cyclol business. It was an aspect of her life and work that was absolutely an empty page to me. I knew her work on crystals, twinning in structures and Fourier transforms, and the things that were really positive that she did — that she made contributions to. It seems to me that whether or not she had a cyclol theory that was good or wasn't is largely a function of the importance that molecular biology has suddenly taken in the last decade, so that people have seized on this part of her work as being significant, whereas there were many other things that she did that were so much more significant.

*Marjorie Senechal:* But she was doing them to verify her cyclol theory.

*Martin Buerger:* She apparently knew that this protein structure was going to be the most important of all those contributions she made, and she underplayed all the others: her contribution to Patterson synthesis and structure factors. She never got any further on structure factors.

*David Harker:* Excellent stuff!

*Mary-Elizabeth Murdock:* In retrospect, that sounds like a mistake.

*Martin Buerger:* She wasn't appraising her work correctly. Is that what you're saying?

*Mary-Elizabeth Murdock:* Yes. Do you think that there was anyone she respected enough to listen to? Could such a person have persuaded her to emphasize something other than the cyclol theory?

*Martin Buerger:* I think she was right in thinking that the cyclol theory was going to be the great theory. If this was correct, it would be the thing that we're feeding on now, but it turned out not to be the case.

*Mary-Elizabeth Murdock:* Didn't she have an inkling of that — or wouldn't she admit it?

*David Harker:* I think her trouble was that as long as she made progress, did things correctly which were accepted, she lost interest in them immediately. Only when she had a battle . . . She fought, unwisely I would say; she would never give an inch, she would never alter her tack.

*Mary-Elizabeth Murdock:* That's rather self-destructive though, isn't it?

*Marjorie Senechal:* She never lost sight of the cyclol as her goal. If you read her papers on twinning in crystals, or her little paper on diffraction patterns in crystallography and astronomy, you'll find that they all begin the same way: "We are engaged in a search for the structure of proteins in which we are attacking this problem from many different points of view; one of the ways is to look at . . . " She seemed to have great breadth of scientific interest and knowledge; she would take different points of view and contribute to other subjects along the way, but the goal was always the protein, as far as I can tell. She seemed to feel that this was her life's work, and all the rest were incidental steps building up to that. That is why the emphasis was the proteins — even though we look back and see contributions made in other areas, to her they were incidental.

*Elizabeth Moore:* To me that's so sad, because she made significant contributions.

*Mary-Elizabeth Murdock:* Well, that is what papers tell us, you see. Once we become fairly familiar with these papers, perhaps we'll get an entirely different perspective on Dorothy Wrinch.

*Marjorie Senechal:* There is something I'd like to ask, since there are people here who can elucidate it. She was urged, according to the papers, to apply herself to mathematics and not to fool around with chemistry anymore. She apparently resented this, but a lot of the work she did *was* mathematical, such as her work on Fourier transforms. So she did find many mathematical contributions to make to protein theory. She wrote that there was a general opposition to her; she felt that she had enemies everywhere. Now in the late 1940's she was invited by John von Neumann to work with him on a protein structure determination project involving the newly-developed computers. She had an enormous contribution to make, and she wrote many letters explaining what she would do and how she would set it up. This is something that all the people who wanted her to do mathematics should have welcomed, yet she couldn't get money for it. She tried to obtain grants to enable her to go to Princeton, and there were many letters written on her behalf, but it came to absolutely nothing. I don't know whether it was the fact that science had changed during the war and was now more bureaucratic, and in these bureaucracies she ran into opposition that was perhaps unthinking, or whether this was something of her own doing. Why wasn't she able to work this out?

*Evans Hayward:* It is unbelievable that if von Neumann really wanted her, he couldn't pay for her. I've read his letter to her and I see what it says, but it is just a contradiction. If he really wanted her, he could have found a way: he had that much power. So I believe he was being polite.

*Marjorie Senechal:* There are three or four more letters that are not on display which go into much more detail. You may be very well right, but he certainly did invite her.

*Evans Hayward:* We listened to a conversation this morning about the buddy system in science. It just happens to be true that if somebody as prestigious as he was wants to do something, he can.

*David Harker:* I think that's right. If it hadn't been for Langmuir, I never could have gotten my protein structure project going.

*Claire Sullivan:* We ought to try to wrap this up now. I will tell you a little bit more about what there is about the Pauling-Wrinch controversy. There are comments made by Pauling and Niemann, sent to Dorothy Wrinch, regarding her article, "The geometrical attack on protein structure", which really was attacking Pauling and Niemann. She received these comments by them, and then she wrote a list of comments on the comments by Pauling and Niemann. Then there was a lot of correspondence with the *Journal of the American Chemical Society,* which tried to play the role of a mediator in the argument. The editor at that time was Arthur Lamb. Here is a letter of 1939 to Dr. Wrinch: "Your manuscript entitled 'The geometrical attack on protein structure' was concerned so largely with the article of Pauling and Niemann published in the *Journal* last year, that it was sent to Dr. Pauling, not as a referee, but for any rebuttal which he might care to make, and have submitted, along with your manuscript, to the referees. A joint rebuttal from Drs. Pauling and Niemann was promptly received. Your manuscript, with this rebuttal, was then submitted to the two referees, accompanied by a letter stating my opinion as to the general policy which the *Journal* should follow in such instances, corresponding substantially I believe, with what I told you in our recent conference, namely, (1) we are under obligation to publish a correction of any demonstrated error of fact or logic in any article which we have published, (2) we are under no obligation to publish any further discussion of a moot point, particularly if such a discussion has already been published elsewhere. Any subject matter, beyond what is required for the adequate correction of the mistakes, must be considered from the standpoint of a new manuscript and to be acceptable must represent a novel and meritorious contribution of a theoretical or experimental nature. Your manuscript and copies of the reports of these two referees were then submitted to two further referees... I have studied these reports with great care and have pondered a good deal on the problem presented by your manuscript. My conclusions pretty much agree with those of the fourth referee. It seems to me that you ought to be permitted to set forth very briefly the demonstrable errors in the article of Pauling and Niemann, interpreted particularly in the light of the present rebuttal of those authors to your manuscript. It would seem to me that it should be possible to do this in about 10 typewritten pages. This new manuscript would then be sent to Pauling and Niemann for their further rebuttal, and this I should expect to be even briefer than your own manuscript — perhaps 5 typewritten pages. This further rebuttal would then be returned to you and, as the referee suggests, the manuscripts would be sent back and forth until both you and Dr. Pauling were content to have them published simultaneously." From referee two: "We're in a mess, and we have to get out of it as best we can. Fortunately, we have the rule against polemics to fall back on. Hence, I feel that there can be no further publication in this vein on the matter in hand. With the publication of the contribution from Dr. Wrinch — to publish which she is clearly entitled in fairness — the subject is closed as far as the *Journal* is concerned, although both parties are, of course, free to contribute new material at any time — provided the pH is not too acid. This reduces the matter to the acceptability of the present Wrinch manuscript. I should prefer to have a little of the phraseology changed — she is a keen debater and bears down rather hard in some places, but she has justification for this. However, to change anything means weeks of delay and time is certainly a factor here — in fairness to Dr. Wrinch. I don't see a thing for us to do but accept the manuscript as it is, and get it into print at once. I think you should also write an editorial letter to both parties and make it perfectly clear that the *Journal* regards the incident as closed. No possible good can come of further public discussion now." And the fourth referee says: "I felt at the time and still feel that the Langmuir-Wrinch and the Pauling-Niemann papers which were published in the *Journal* definitely warranted publication. This was not because of any conviction that the ideas presented in either paper were correct, but merely becuse both papers prescnted new ideas of importance to protein chemistry. Progress can be made in research only by free and open presentation and discussion of new ideas and it is always quite possible that any idea which later may turn out to be completely wrong may still have been of the greatest importance if the discussion of this incorrect idea serves to stimulate thought and perhaps bring out the correct answer. The present paper by Dr. Wrinch seems to be a summary of previous work, a rebuttal to the Pauling-Niemann criticism and a presentation of some further ideas. Further speculation seems unjustified at present and recapitulations are not necessary. It does seem that Dr. Wrinch is entitled to a rebuttal to the Pauling-Niemann paper. I definitely can not see the necessity for any great rush. I would favor a modified procedure, embodying the conditions laid down in your letter and in that of Pauling, but not quite as rigid as suggested by referee #1. The present paper is not suitable for publication. Anything definite and specific that is worth saying could be presented not only as well but far more effectively in much less than twenty pages. I suggest that the paper be returned to Dr. Wrinch with your statement of the situation in referee #1's remarks and criticism. If she still desires to publish a rebuttal to Pauling and Niemann, it should be prepared in concise form and so far as possible along the lines suggested by referee #1 with reference to previous arguments where absolutely imperative but with no further speculation. This paper would be sent to Pauling for preparation of his rebuttal. Both papers would then go back and forth until both parties were satisfied with dropping the matter on simultaneous publication of the two papers. My proposal is essentially referee #1's proposal slightly relaxed. This seems to me to be perfectly fair to Dr. Wrinch. The proposed two papers will both be a debate on the structure of proteins. This is simply an opportunity for Dr. Wrinch to answer any objections or point out any incorrect statements in the Pauling-Niemann paper. If she goes at this in a fair and reasonable way she should not need to bring in very much from previous publications. I do not agree with the referee that haste is imperative. It is probably true that both parties are guilty of being excessive in their statements. The fairly long pause which will probably come while the two papers are passed back and forth may be exactly what is needed to give each a chance for cooling down and taking second thoughts. I think a lot of wars could have been stopped that way; too bad something of the sort could not have been used in Europe." So apparently this was quite a battle, which was not seen in the final publications; by just reading the journal you would have no idea what was actually behind getting those two papers published.

*David Harker:* I remember when we were writing the paper on bond energies and she wanted to put in, "Anyone, with an interest in elementary arithmetic, can see..." That kind of thing, but I wouldn't let her.

*Martin Buerger:* Are her monograph manuscripts available?

*Claire Sullivan:* Yes.

*Martin Buerger:* I imagine if you looked them over, you would find very few errors in the original copies.

*Marjorie Senechal:* I really do think that this is going to be a treasure for years to come, with the kinds of things that are in here and the kinds of insights they provide.

*Claire Sullivan:* Not only is it very valuable for anybody studying the history of the cyclol theory or her actual work, but it also tells a lot about how science is done, applying for grants, acceptances, rejections, the amount of work going into getting a paper published, and the models that illustrate the theory that eventually gets published — you normally see only the final publication which leaves out a lot. And also the amount of correspondence that went back and forth between her and many, many scientists tells a lot about how science works. She was not just sitting alone at her desk, coming up with all her theories, it involved a lot more. And also it's a biography of a very prominent woman scientist and of a fascinating person as well. So from a purely historical point of view or biographical point of view, it's very valuable.

# Papers and Books By Dorothy Wrinch

**1919** On some aspects of the theory of probability. *Phil. Mag.* 38, 1919, 715-731 (with Dr. H. Jeffreys).

On the exponentiation of well-ordered series. *Proc. Camb. Phil. Soc.* 19, 1919, 219-233.

**1920** On the structure of scientific inquiry. *Proc. Aristotelian Soc.* 1920-1921, 181-210.

**1921** The relation of geometry to Einstein's theory of gravitation. *Nature* 106, 1921, 806-809 (with Dr. H. Jeffreys).

On certain fundamental principles of scientific inquiry. *Phil. Mag.* 42, 1921, 369-390 (with Dr. H. Jeffreys).

Relativity, International Congress of Philosophers. Paris, 1921.

A generalized hypergeometric function with n parameters. *Phil. Mag.* xli, 1921.

An asymptotic formula for the hypergeometric function $_0 \, _1(z)$. *Phil. Mag.* xli, 1921.

**1922** The theory of relativity in relation to scientific method. *Nature* 23 March 1922.

On certain methodological aspects of the theory of relativity. *Mind* October 31NS, 1922, 200-204.

The idealistic interpretation of Einstein's theory. *Proc. Aristotelian Soc.* 1922-1923, (symposium with Professor Carr, Professor Nunn, Professor Whitehead and Lord Haldane).

On the lateral vibrations of bars of a conical type. *Proc. Roy. Soc.* A. 101, 1922, 493-508.

On the orbits in the field of a doublet. *Phil. Mag.* 43, 1922, 993-1014.

On the rotations of slightly elastic bodies. *Phil. Mag.* xliv, 1922.

**1923** On the seismic waves from the Oppau explosion of 1921, with Harold Jeffreys, Sept. 21, *R.A.S. Monthly Notices* Jan. 1923.

On certain aspects of scientific thought. *Proc. Aristotelian Soc.* 1923-24, 37-54.

On certain fundamental principles of scientific inquiry (Second Paper) *Phil. Mag.* 45, 1923, 368-374 (with Dr. H. Jeffreys).

The theory of mensuration. *Phil. Mag.* 46, 1923, 1-22 (with Dr. H. Jeffreys).

On mediate cardinals. *Amer. J. Math.* 45, 1923, 87-92.

On the lateral vibrations of rods of variable cross-section. *Phil. Mag.* 46, 1923, 273-290.

Some approximations to hypergeometric functions. *Phil. Mag.* xlv, 1923.

Table of the Bessel function $I_n(x)$, with Hugh E.H. Wrinch. *Phil. Mag.* xlv, 1923.

**1924** Scientific thought. *Mind*, 33NS, 1924, 184-192.

Some problems of two-dimensional electrostatics. *Phil. Mag.* 48, 1924, 692-703.

The hypergeometric function with k denominators. *Quart. J. Math.* 50, 1924, 204-224.

Some boundary problems of mathematical physics. *Proc. Lond. Math. Soc.* 24, 1924, 435-458.

Some problems of two-dimensional hydrodynamics. *Phil. Mag.* 48, 1924, 1089-1104.

**1925** Fluid circulation round cylindrical obstacles. *Phil. Mag.* 49, 1925, 240-250.

On the electric capacity of certain solids of revolution. *Phil. Mag.* 50, 1925, 61-70.

Laplace's equation and the inversion of co-ordinates. *Phil. Mag.* 50, 1925, 1049-1058.

Laplace's equation and surfaces of revolution, with J.W. Nicholson, *Proc. Roy. Soc.* A. 108, 93-104, 1925.

**1926** On the pressure distribution round certain aerofoils of high aspect ratio. *J. Roy. Aero. Soc.* 30, 1926, 129-142.

The roots of hypergeometric functions with a numerator and four denominators. With H.E.H. Wrinch. *Phil. Mag.* vol i, 1926.

Scientific methodology with special reference to electron theory. *Proc. Aristotelian Soc.* 1926-27, 41-60.

**1927** The relations of science and philosophy. *J. of Philosophical Studies* 2, 1927, 153-166.

Electrostatic problems concerning certain inverted spheroids. *Phil. Mag.* 3, 1927, 865-883.

A class of integral equations occurring in physics. *Phil. Mag.* vol. iv, 1927, with J.W. Nicholson.

**1928** Aspects of scientific method. *Proc. Aristotelian Soc.* 1928-29, 94-122.

On the asymptotic evaluation of functions defined by contour integrals. *Amer. J. Math.* 50, 1928, 269-302.

On spheroidal harmonics as hypergeometric functions. *Phil. Mag.* 1928, 6, 1117-1122.

On a method for constructing harmonics for surfaces of revolution. Proc. International Congress of Mathematics, Bologna, 1938.

**1929** Scientific method in some embryonic sciences. *Proc. Aristotelian Soc.* 1929-30, 229-242.

On the structure of serial relations. *Phil. Mag.* 1929, 8, 698-702.

On the multiplication of serial relations. *Phil. Mag.* 1929, 8, 1025-1042.

**1930** On harmonics applicable to surfaces of revolution. *Amer. J. of Math.* 1930, 52, 305-318.

On some integrals involving Legendre polynomials. *Phil. Mag.* 10, 1037-1043, 1930.

*The Retreat From Parenthood*, (pseudonym: Jean Ayling), Kegan Paul, Trench, Trubner & Co., Ltd., 1930.

**1932** Some problems concerned with inverted prolate spheroids. *Phil. Mag.* 14, 1061-1078, 1932.

Harmonics associated with certain inverted spheroids. Proc. International Congress of Mathematics, Zurich, 1932.

Applications of prolate spheroidal harmonics. *Phil. Mag.* Ser. 7, vol. xiv, 1932, pp. 829-848.

**1934** Chromosome behavior in terms of protein pattern. *Nature* 134, 978, 1934.

**1935** The contractile factors of the chromosome micelle. *Nature* 134, 788, 1935.

The chromosome micelle and the banded structure of chromosomes in the salivary gland. *Nature* 136, 68, 1935.

Origin and value of egg case in the crustacea. *Nature* 136, 68, 1935.

Chromosomes and molecular aggregates. Internations Congress of Botany, Amsterdam, 2, 24, 1935.

**1936** On the molecular structure of chromosomes. *Protoplasma* 25, 550, 1936.

The pattern of proteins. *Nature* 137, 411, 1936.

Energy of formation of cyclol molecules. *Nature* 138, 241, 1936.

Structure of proteins and of certain physiologically active compounds. *Nature* 138, 651, 1936.

The hydrogen bond and the structure of proteins (with D. Jordan Lloyd), *Nature* 138, 758, 1936.

**1937** Built-up films of proteins and their properties (with Irving Langmuir and V.J. Schefer), *Science* 85, 76, 1937.

Nature of the linkages in proteins. *Nature* 139, 718, 1937.

Intramolecular folding of proteins by Keto-Enol Interchange (with W.T. Astbury), *Nature* 139, 798, 1937.

On the pattern of proteins. *Proc. Royal Society* 160A, 59, 1937.

The cyclol theory and the globular proteins. *Nature* 139, 972, 1937.

On the structure of insulin. *Science* 85, 566, 1937.

The cyclol hypothesis and the globular proteins. *Proc. Royal Society* 161A, 505, 1937.

On the structure of insulin. *Trans. Faraday Society* 33, 1368, 1937.

On the structure of pepsin. *Phil. Mag.* 24, 940, 1937.

The structure of proteins and its biological significance. International Congress of Physics, Chemistry and Biology, Paris, 395, 1937.

**1938** Structures proposed for protein molecules. *Proc. Physical Society,* 50, 141, 1938.

On the hydration and denaturation of proteins. *Phil. Mag.* 25, 705, 1938.

The structure of the insulin molecule. *Science* 88, 148, 1938.

The structure of insulin. *J. American Chemical Society* 60, 2005, 1938.

The molecular weights of the globular proteins. *Phil. Mag.* 26, 313, 1938.

The structure of the insulin molecule (with Irving Langmuir). *J. American Chemical Society* 60, 2247, 1938.

Is there a protein fabric? Cold Spring Harbor Symposium 6, 122, 1938.

Vector maps and crystal analysis (with Irving Langmuir). *Nature* 142, 581, 1938.

Crystal analysis and point sets. *Nature* 142, 955, 1938.

**1939** The geometry of discrete vector maps. *Phil. Mag.* 27, 98, 1939.

Nature of the cyclol bond (with Irving Langmuir). *Nature* 143, 49, 1939.

The structure of the globular proteins. *Nature* 143, 482, 1939.

The analysis of discrete vector maps. *Phil. Mag.* 27, 490, 1939.

Note on resonance in proteins. Cold Spring Harbor Symposium 7, 48, 1939.

The geometrical attack on the problem of protein structure. *Proc. Royal Society* 127B, 24, 1939.

The structure of insulin and the cyclol hypothesis. *Nature* 143, 673, 1939.

A note on the structure of insulin (with Irving Langmuir). *Proc. Physical Society* 51, 613, 1939.

The tuberculin protein molecule. *Nature* 144, 77, 1939.

Geometry in the service of crystal physics. Physical Soc. Exhibition Jan. 1939.

**1940** The structure of proteins with special reference to cytogenetica. *J. of Genetics*, 40, 359, 1940

The cyclol hypothesis. *Nature* 145, 669, 1940.

Lengths and strengths of atomic bonds (with David Harker), *J. Chem. Phys.* 8, 502, 1940.

The fabric theory of protein structure. *Phil. Mag.* 30, 64, 1940.

The Patterson projection of the skeleton of the structure proposed for the insulin molecule. *Nature* 145, 1018, 1940.

A new calculation of the C=C bond energy and of resonance in certain molecules. *Science* 92, 79, 1940.

**1941** The geometrical attack on protein structure. *J. American Chem. Soc.* 63, 330, 1941.

Further developments of the fabric theory of protein structure. *Phil. Mag.* 31, 177, 1941.

The native protein theory of the structure of cytoplasm. Cold Spring Harbor Symposium, 9, 1941.

Proteins in action, *Collecting Net*, 16, 121, 1941.

Further implications of flexible protein frameworks, *Collecting Net*, 16, 177, 1941.

**1942** The structure of biologically active membranes. *Collecting Net*, 17, 83, 1942.

Native proteins, flexible frameworks and cytoplasmic organization. *Nature* 150, 270, 1942.

The structure of biologically active membranes. *Biological Bull.* Vol. 83, no. 2, Oct. 1942.

**1943** Growth and form. *Isis* 34, 232, 1943.

Review of Growth and form by D'Arcy Thompson, History of Science Soc., 1943.

**1944** Native protein crystallography and diffraction patterns. *Biological Bulletin* 7, 157, 1944.

**1945** Fourier transforms and structure factors. *Phys. Rev.* 67, 198, 1945.

A tetrahedral framework for native proteins? *Biological Bulletin* 89, 192, 1945.

**1946** *Fourier Transforms and Structure Factors*, published by the American Society for X-ray and Electron Diffraction, January 1946.

Patterson distributions and native protein crystallography. *Nature* 157, 226, 1946.

The nature of drug action. *Australian Journal of Science* 8, 103, 1946.

Synthetic Patterson maps (with A.D. Booth), *Journal of Chemical Phys.* 14, 503, 1946.

Biological specificity and the synthesis of native proteins. *Collecting Net*, September 1946.

Active patches on native proteins. *Australian J. of Science*, 8, 103, 1946.

On the nature of biological specificity. *Biological Bulletin* October 1946.

Biological specificity and the synthesis of native proteins. *Biological Bulletin* October 1946.

**1947** Twinnings and intergrowths. *American Mineralogist* 32, 695, 1947.

Patterson-Harker maps (with A.D. Booth). *Journal of Chemical Physics*.

Patterson-Harker maps of molecular crystals. *Journal of Chemical Physics* vol. 15, no. 6, June 1947.

The native protein. *Science* 106, 73, 1947.

Native proteins and biological morphology. *Biological Bulletin* 93, 202, 1947.

Proteins as polycondensations of amino acids. *Bull. of Am. Phys. Soc.* 22, 13, 1947.

Protein synthesis. *Phil. Mag.* 38, 373, 1947.

Native proteins and biological morphology. *Biological Bulletin* 93, 202, 1947.

**1948** The protein: a physicochemical individual. Wallerstein Laboratories Communications 11, 175, 1948.

On the relation between certain distributions and their corresponding Patterson distributions. *Phil. Mag.* 39, 692, 1948.

A note on the structure of insulin. *Journal of Chemical Physics* 16, 1007, 1948.

Biological specificity and protein structure. *Biological Bulletin* 95, 247, 1948.

The cage hypothesis and a common feature of X-ray diffraction studies of proteins. *Biological Bulletin* 95, 272, 1948.

An application of Fourier transforms to a crystal structure analysis. *Am. Mineral.* 33, 782, 1948.

The native proteins as polycondensations of amino acids. *Science* vol. 107, no. 2783, April 1948.

Some crystalline hemoglobins. *Amer. Mineral.* 33, 781, 1948.

**1949** Radial distributions in Patterson space (with A.D. Booth) *Nature* 163, 169, 1949.

The structure of insulin and the cyclol hypothesis. *Biological Bulletin* vol. 97, No. 2, Oct. 1949.

The particle status of proteins. *Biological Bulletin* vol. 97, no. 2, Oct. 1949.

Biological specificity and protein structure. *The Collecting Net*, vol. 29, no. 1, Nov. 1949.

Diamond networks in reciprocal space and Fourier transforms of cyclol structures. Meeting of the American Society for X-ray and Electron Diffraction, Dec. 1949.

**1950**  A note on the diffraction patterns of proteins. *Acta Cryst.* 3, 76, 1950.

Patterson-Harker maps of protein crystals. *Amer. Mineral.* 35, 124, 1950.

Water and biological function. *Biological Bulletin* vol. 99, no. 2, Oct. 1950.

Water and biological morphology. *Biological Bulletin* 1950.

The 10-11A spacing of crystalline proteins and the cyclol hypothesis. Meeting of Am. Cryst. Assoc. April 11, 1950.

Vector functions for uniform spheres and spherical shells. *Phys. Rev.* 79, 203, 1950.

Certain Fourier transforms and structures of proteins. *J. Chem. Phys.* 18, 562, 1950.

Vector maps of crystals containing particles. *Am. Cryst. Soc.* Aug. 1950.

Vector maps of hydrated protein crystals. *Acta Cryst.* 3, 475, 1950.

**1951**  Multiple cube systems in mineralogy. Meeting of Am. Cryst. Assoc. Feb. 1951.

Some indications regarding protein structure from X-ray studies. Inter. Cryst. Congress, Stockholm 1951.

Some remarks on Fourier transforms and vector maps in structure determinations. Symposium on Structure Determinations, Stockholm 1951.

Water in protein crystals. *Biological Bulletin* vol. 101, no. 2, Oct. 1951.

Protein molecules with relatively low molecular weights? *Biological Bulletin* vol. 101, no. 2, Oct. 1951.

**1952**  Fourier transforms and intensities from crystalline proteins. *Bulletin Am. Phys. Society*, vol. 27, #1, p. 36, Jan. 1952.

The bearing on new results in protein structure on cytological interpretations, *Biological Bulletin*, 1952.

Fourier transforms and intensities from protein crystals. *Phys. Rev.* 86, 611, 1952.

Skeletal units in protein crystals. *Science* 115, 356, 1952.

The twinning of cryolite. *Amer. Mineral.* 37, 234, 1952.

Indications regarding protein structure from protein crystallography. Symposium on Protein Structure, abstract 1. Meeting of Am. Cryst. Assoc. June 1952.

Some observations on twinning. Symposium on Twinning, abstract 1. Meeting of Am. Cryst. Assoc. June 1952.

On the vector map of crystalline horse hemoglobin. *J. Chem. Phys.* 20, 1051, 1952.

A megamolecular structure factor. *Phil. Mag.* 43, 801, 1952.

Evidence for globulite molecules in horse hemoglobin. *J. Chem. Phys.* 20, 1332, 1952.

Globulite units in protein crystals. *Acta Cryst.* 5, 694, 1952.

Pseudosymmetry of crystals and atomic patterns. Meeting of Mineral. Soc. of Am. Nov. 1952.

Molecules of the insulin structure. *Science* 116, 562, 1952.

Recent results regarding the structures of the globular proteins. *Biological Bulletin*, vol. 103, no. 2, 313, Oct. 1952.

**1953** Fourier transforms of shells and intensities from crystalline proteins. *Phys. Rev.* 91, 471, 1953.

Proposed polypeptide chain configurations and the structure of horse hemoglobin. *Acta Cryst.* 6, 562, 1953.

The hypothesis of parallel rodlike polypeptide chains in horse hemoglobin. *Acta Cryst.* 6, 638, 1953.

The structure of ribonuclease. *Biological Bulletin* 105, 423, 1953.

A globulite structure for acid insulin sulphate? *Biological Bulletin* 105, 456, 1953.

Evidence for globulite molecules in ribonuclease. *J. Chem. Phys.* 21, 2099, 1953.

Diffraction patterns in 19th Century Astronomy and 20th Century crystallography. From a collection of essays in honor of Charles Singer entitled "Science, Medicine and History," vol. 2, p. 197, Oxford Univ. Press 1953.

Nomenclature of cyclohexane bonds. *Nature* Dec. 1953.

**1954** Vector maps of some distributions in continuous space. Am. Cryst. Assoc. Apr. 1954.

The structure of horse hemoglobin in the light of the intensity map of the horse methemoglobin crystal. *Acta Cryst.* 7, 353, 1954.

On the interpretation of vector maps. Intern. Congress Cryst. Paris. *Acta Cryst.* 7, 627, 1954.

On the interpretation of vector maps — abstract.

Native protein structure and a tetrahedral motif. *Biological Bulletin* 107, 323, 1954.

**1955** Native protein structure in the light of physico-chemical findings. *Exper. Med. and Surgery* 13, 33, 1955.

Vector distributions and crystalline proteins. *J. Chem. Phys.* 23, 1361, 1955.

On the meaning of the vector map of horse methemoglobin. *Acta. Cryst.* 8, 515, 1955.

On the structure of DNA (39 pages, unpublished).

**1956** Diffraction star produced by magnetite within muscovite (with B.M. Shaub), *Amer. Mineral.* 41, 744, 1956.

Notes on six-rayed diffraction star produced by magnetite enclosed in muscovite (with B.M. Shaub), *Amer. Mineralogist* 41, 994-947, 1956.

**1957** The dark ages' scourge led to lifesaving drugs. *New York Herald Tribune*, July 21, 1957. Reprinted in the *Smith Alumnae Quarterly*, February 1958.

The structure of bacitracin A. *Nature* 179, 536, 1957.

An approach to the synthesis of polycyclic peptides. *Nature* 180, 502, 1957.

**1960** *Chemical Aspects of the Structure of Small Peptides*. Munksgaard, Copenhagen, 1960.

**1962** Some issues in molecular biology and recent advances in the organic chemistry of small peptides. *Nature* 193, 245, 1962.

**1963** Recent advances in cyclol chemistry. *Nature* 199, 564, 1963.

**1965** *Chemical Aspects of Polypeptide Chain Structure and the Cyclol Theory*, Munksgaard, Copenhagen, 1965; Plenum Press, New York, 1965.

A contemporary picture of the chemical aspects of polypeptide chain structures and certain problems of molecular biology. *Nature*, vol. 206, pp. 459-461, 1965.

BALTIMORE.

Dear Dr Pauling

Your attacks on my mother have been made rather too frequently. If you both think each other is wrong it is best to prove it in stead of writing disagreeable things about each other in papers. I think it would be best to have it out and see which one of you is really right. There are many quarrells in the world Alas!! Dont please let yours be one it is ~~these~~ these things that help to make the world a Kingdom of misery!!

Yours

Marian P. Wrinch

The Victoria
UNIVERSITY OF MANCHESTER
Department of Chemistry
23rd, May, 1935.
Professor M. Polanyi

Dear Dorothy,

May I try to sum up the position as I think people see it here now, with regard to your work.

1. That the chromosome must have a fibre structure since the organism developed from it is fibrous is certainly a fundamental and true idea.

2. To connect this fibre structure with the genetic functions is an equally good idea.

3. To attribute to the side chains the genetic identity of the chromosomes is also a brilliant idea, but one cannot feel just as sure that it is true as with respect to 1 and 2.

4. To take some sort of longitudinal structure linked up by some sort of annular structure is obviously a trustworthy leading idea; though tentative to some extent.

5. The clupein and nucleic acid idea I should think is a brilliant demonstration of the workability of the whole scheme, but it might drop out of the picture altogether in future without impairing its essential value, just as the glycogen-lactic acid model has dropped out, — ten years after its author got the Nobel prize for it — out of the Meyerhof mechanism to be substituted by the glyco phosphate model leaving the fundamental discovery of Meyerhof untouched.

6. Swelling as a basis of the splitting of chromosomes is certainly something to hold on to; but remember the various models of amoeboid motion (Rumbler) based on surface tension which never became useful.

When I write this down, I am under the impression of a talk with Ritchie who, I think, has the most intelligent view on the matter and is if anything, rather more emphatic on the good points than I have been. Speaking as a distant relative of the gypsies, I might add, as a piece of fortune-telling, that you will live to great recognition of your vision.

Best wishes,
Yours,
M.P.

THE UNIVERSITY
EDGBASTON,
BIRMINGHAM 15
*STRICTLY PERSONAL AND CONFIDENTIAL*       8th May, 1942.

Dear Dorothy,

Many thanks for your reprints. I had been intending to write to you in strictest confidence about a matter which you will see must be treated *extremely* confidentially. Between you and me and nobody else, this year I am again on my Sectional Committee of the Royal Society and am giving the Croonian Lecture. It is not impossible that I shall be on the Council in the near future. You will appreciate that I should be very glad to take any *tactful* steps to promote the successful candidature of a woman for the Royal Society. I think you know some of the snags in the past, and you will appreciate that in promoting such an issue for the first time it is important to sponsor a really outstanding person. I cannot honestly say that I think there is a woman candidate in pure biology of such outstanding merit as to silence all plausible antagonism. On the other hand I might do more harm than good by erring from my proper beat. What this means is that I might take tactful steps to catalyze your candidature if I knew exactly what to do. You will appreciate that this does require very tactful handling, if you approve and have not been approached from other quarters. On that assumption I suggest that you might give me two or three names of persons who would not resent a tentative suggestion if I were to write to them in a personal and friendly vein. You will also appreciate the necessity of confining any action I might take to persons who are not likely to be antagonised by the intervention of someone who is not himself a professional mathematician. Meanwhile all good wishes.

Ever yours,
Lancelot Hogben

P.S. If you could get Enid an invite to give a lecture at Amherst on vital statistics of some sort you could discuss the matter with her and transmit your views without compromising yourself by writing a letter as irregular as this.

· · · · ·

MBL WOODS HOLE.   May 26, 1942

My dear Lancelot, Your letter was very nice to get. I was so glad to hear from Julian Huxley, when he was here, that you were pleased to be in Bir. and that you were happier than in A. My best wishes for the work. We have also been hearing about your book from Muller. I hope it goes very well.

The war is enveloping everything here. A. in particular is very much on a war basis, what with a complete new semester put in between June and Sept. and many people being fired, right and left. I don't know whether the authorities are correct in thinking that there will be a bad drop in the number of students, but anyway they are trimming their sails for this. Lecs of all kinds from outside are being cut to pieces, so I don't see how it will be poss. to get an invitation for E. But I shall keep the eyes open to see if it can be managed. I wish she would send me a line sometimes.

With regard to the other topic in your letter ... My supporters who understand something of my objectives, even if little or nothing about the methods are D'Arcy T. and Donnan and, oddly enough, JBS. and also William Wilson, physics Bedford coll. conceivably tho not very likely RHFowler. I think Bragg fils a definite enemy. Conceivably Dale, and very likely Rosenheim Nat Inst Med Res, also Kennaway Cancer free hop Fulham road Astbury very definitely an enemy, also Desmond (alas alas). It seems to me the line that my stuff if right will be quite something is worth pressing (if I may make so bold as to say so). If wrong, it won't be the first time that this has happened even to Nobelists (cp sterols). My really strong suit is of course Irving Langmuir (Res. Lab. General Electric Co. Schenectady NY). He is a foreign member I imagine. Also Harold Urey (Columbia Univ NY NY) who has been helpful. Nils Bohr would be an ideal supporter and would be as good or better than Irving, if he can be reached in Copenhagen. Another possibility is Albert Einstein (Inst. for Advanced Study Princeton) who is interested and discusses knotty protein points with me on occasion.

As to the field, I definitely think it should be biology or chemistry and not mathematics. Molecular biology would be the best name I think.

I think this is all very good of you and I write quickly and maybe over frankly, cos of life being so short. I shall not attempt to hide from you the fact that I do think my work is turning up trumps and also that this plan which you hint at would be the greatest single factor that I can think of towards fostering my work in the future. I don't intend to let it die except over my dead body, though all the diehards spend more time than they should trying to kill it. On the other hand, something on the positive side, such as this, would smooth my path quite tremendously in the future.

Once again, dearest Lancelot, don't attempt anything via the mathns. I am hopelessly compromised by applying the noble art to living material.

My best wishes and affectionate greetings. I hope your progeny is flourishing. I wish so much you were coming over.

* * * * *

from  
   Professor M. Polanyi, FRS  
   Telephone: ARDwick 2681

The University of Manchester  
Department of Social Studies  
24.3.48

My dear Dorothy

I was happy with your letter showing that you had not forgotten us. You and I have much in common in the manner we managed to make our scientific career less dull than usual. Your fidelity to your studies of proteins is, however, unmatched by any of my own passions. I much admire you for your steadfastness. I shall send you with pleasure some of my writings. There is a pamphlet on the Foundations of Academic Freedom that I have available, but unfortunately I do not know whether this kind of thing does not make you sick. It has that effect on many people I know. Throw it away if it makes you feel that way before it has caused any permanent damage to your insides or to our friendship.

I hope you are happy in America. The feature of their great civilization which I admire most is their eloquence. I have just put down J.R. Oppenheimer's paper Physics in the Contemporary World in the Atomic Scientists Bulletin. What splendid prose!

We have just moved to a new house further out in the country and I fished out some of your old letters on this occasion. I was reminded of having been a little puzzled by the book which you had forgotten at our house, "This Bed Thy Centre" — later I discovered the line in Donne's great poem. Magda asks me to send you her love with mine. We both hope to hear from you again when you think of us.

Michael

· · · · ·

THE INSTITUTE FOR ADVANCED STUDY
School of Mathematics
Princeton, New Jersey
December 24, 1946

Dr. Dorothy Wrinch
81 Woodside Ave.
Amherst, Mass.

My dear Miss Wrinch,

I am returning with the same mail some of your reprints, which you were so good to loan to me, marked "last copy, please return". I want to thank you for having given me the opportunity to read them.

I would also like to thank you most cordially for your visit to Princeton, I learned a great deal from you, and I only hope that our exchanges will continue in the future.

In this connection I would like to mention, that I saw Dr. I. Langmuir at a biophysical conference in New York, the day after we met. I mentioned to him that I thought that protein molecule models might be tested against x-ray diffraction patterns by diffracting centimeter shortwaves on metal models scaled up by factors of the order $10^8$. He showed a good deal of interest in this, and he told me that D. Harker was actually working at the General Electric laboratories in Schenectady with parallel cm wave beams, with a definite idea towards x-ray crystallographical applications. We agreed that I would visit Harker at Schenectady sometime in January and discuss these things with him and Langmuir. I think that it would be most desirable, that you, too, should participate in these discussions.

I also talked with Dr. Zworykin of the RCA Laboratories in Princeton. I found him also very responsive to the idea of making "scaled up" experiments to test molecular models with cm shortwave in order to compare the results with the observed x-ray diffraction patterns. He offered to develop most of the instrumentation that will be required, in particular for scanning and recording, and he is also willing to provide the personnel for this work.

Some radar items may have to be obtained on loan from Government Surplus stocks, but I think that I can arrange for that. We will also have to provide crystallographical and chemical guidance and the actual models. All of this would, of course, depend quite vitally on the possibility of a continuous cooperation with you. May we count on that? I think that I can make all the necessary arrangements for your visits, for expenses in connection with making models, etc. Of course, we should have some preliminary conferences, before the real work begins.

These things will get under way somewhat slowly, since I will be away on a vacation from December 30 to January 14, and since a number of different people will have to be "coordinated". I hope, however, that by February things will be moving reasonably fast. I would appreciate, on the other hand, your reactions to these plans and your advice on any phase in them, as soon as possible — if feasible before I leave. Hoping to hear from you soon, I am,

Sincerely yours,<br>John von Neumann

* * * * *

81 Woodside
Amherst, Mass.
Dec. 27, 1946.

Dear Professor Shapley,

In continuation of our conversation at the MBL in the summer, I would like to report to you the following development. I would also be very happy to hear of the outcome of the conversation which you were planning to have in Boston at the end of Sept. last. Did it prove possible to interest Prof. Conant and were any plans formulated in that or any other direction?

The development is as follows: Apparently Irving Langmuir met John von Neumann and actually interested him so much in the protein project that he asked me to go down and discuss it with him in Princeton. This I did. I found that the major outlay contemplated in our budget for the projected 5 year protein scheme could be eliminated out right, since he proposed that his electronic computer, now under construction at Princeton backed by large funds, should be used for the protein work. There is actually no one whom I would have wanted to interest in the protein scheme more than vN — indeed, as he reminded me, I tried to do just this already in 1938 — since his mathematical genius and his wide physical interests are just what is really needed for a successful outcome. It emerged in our conversations that he is really interested in the protein, not only in the electronic computer side of the project. We made a number of plans as to the form of the project — the techniques etc. — and I heard this week that these are now going ahead with very considerable cooperation promised by RCA at Princeton — instruments to be developed and personnel which will be provided — and also by Irving who is proposing to bring my good friend David Harker, also at GE, into the scheme. This last letter from vN asks if he can count on my continuous cooperation. I quote . . . 'All of this would, of course, depend quite vitally on the possibility of a continuous cooperation with you. May we count on that? . . . We will have to provide crystallographical and chemical guidance. I think I can make arrangements for your visits, for expenses in connexion with making models, etc. Of course, we should have some preliminary conferences, before the real work begins. . .'

I give all these data so that you may see the present situation. Please could you now advise me on the one difficulty which remains. I have replied to vN that I will certainly cooperate. But what I want is to transfer my work bodily to Princeton and to find some academic niche there — if possible including the task of teaching crystallography to chemists etc. — indeed to make a place for myself there and press ahead with my whole protein work there, where there would be a chance of scientific intercourse and collaboration with mathematicians at the Institute for Adv. Stud. with chemists at the University and with their biologists, to say nothing of RCA etc. Is there any fund, e.g. your Science Fund or any other to whom I can now turn to provide some or all of the money needed. I feel that if I could obtain some of it in this way, a place might be found for me in one of these departments there. And there is the very important factor now, which was not there when we talked, namely the fact that vN has now invited me for 'continuous collaboration' which I take it carries with it his backing in my quest for funds. This is of course a personal letter and I am venturing to ask you to direct me in this quest, since I really don't yet know my way about in my new country. Also there is this ever present opposition of which you know — which must be circumvented if I am ever to have my chance to show what I can do — (for it must be remembered that I have as yet been given no such chance at all, and frankly I now feel that I could get a long way to the solution of the problem.) You may remember in the plan I showed you there were two people only, involved, myself and A.D. Booth of London plus assistants. With vN's computer now to be available, some of the assistants also are no longer needed. Furthermore I have just heard from Booth that the Rockefeller Foundation has agreed to send him to vN for a year from March next and he is coming all complete with an assistant provided by the Rubber Res. Ass. of London I think. Thus the whole thing winnows down to me plus an assistant for the 5 years, since I think that Booth will manage the other 4 years from the RF and in any case, not being an American citizen, he would not be eligible for American science funds, except for RF and a few others. So this is the situation. Our original budget, in the scheme discussed with IL, was in 6 figures and it is now 5 times $(A+B)$ — A for me, B for the assistant.

Please forgive me for this long letter — and please advise me.

Sincerely yours,<br>Dorothy Wrinch

· · · · ·

National Science Foundation<br>Biological and Medical Sciences<br>Proposal Evaluation Sheet

In 1936 Dr. Dorothy Wrinch introduced the concept of cyclol groupings in proteins and polypeptides, and ever since she has bravely and courageously defended her theory against the ensuing attacks of her opponents that often went beyond the lines of objective criticisms. In recent years striking results were obtained in the laboratories of many distinguished chemists strongly suggesting that some of the structural problems pertaining to polypeptides and other types of organic compounds may be resolved by the introduction of multiple peptide or cyclol bonds and groupings to supplement the normal peptide groupings. The time has come when all the pertinent information scattered in the literature and collected by a number of European and American scientists must be assembled, evaluated and integrated with special reference to Dr. Wrinch's cyclol theory. In my opinion, this important work of fundamental nature can best be helped by granting Dr. Wrinch's modest request for assistance as submitted in her highly meritorious proposal.

Eugene Pacsu
1958
*A cherished document!*
*DW*

. . . . .

Dearest Erice:

Now that good letter (undated) was very much to the point. Now I have something specific and I really have hopes that I can improve — of course I just love the assoluta . . . and when shall we see a ballet again? Thank you so much for this most constructive and helpful criticism — instead of vague and broad statements which just worry me. But please Erice don't tease me by laughingly saying you can't put that sheet of paper which I specially asked you to return into the airmail letter and never returning that second copy, which I typed out, at all, at all, as the Irish say. If you realized that the MonOne is, in my scale of reckoning, a digit against something in the hundreds of thousands for ProtMon you would not do this to me. And please keep a list of expenditures cos it irks me so much to know of all my debts to you and not have some record somewhere. After all with almost everything you do for me such that I can never repay you, just please keep a record of those which I can and will. Oh dear, I am low in spirit about the ProtMon and I suppose I must just try a bit harder. By the way, there were a few remarks here and there in your summer essayettes making comments about how to use words in all these little epistemological issues and I am preparing all the time to get these matters dealt with adequately in the ProtMon even if only in an appendix. However the vocab. is a real difficulty and that is why you will find this article enclosed (keep this copy it is for you). I don't know what you will think of it but I know of nothing better except what you yourself could write on the subject. So do please tell me how you think his vocab should be modified and how much of it we can use. It gets a bit tiresome when one (I) always seem to want to use words in a way different from anyone else, so see how much of this we can standardize for our use as there are bits of this article that I shall want to quote one of these days. O dear Erice — such a dark time this week: we have an awful lot of the big brass coming round to give us lecs for which they are paid large sums and this week there was a fellow who talked such frightful dreadful rubbish about proteins, this not being his field at all, en route to discussing his own experiments — cancer res. — which may be of importance. I cannot understand what has got into so many of these so-so-so successful people. They pursue the common course of repeating what they have been told to think about proteins, take no responsibility for it, do not discuss evidence for what they say, are, so far as I can see, taking the line that what is good enough for Linus Pauling is good enough for them, and if challenged to give chapter and verse for what they are taking as fact would certainly want to say 'But why for Christ's sake pick on me, I am not a protein man, I am just one person and there are thousands and thousands of us and we all say the same, and in any case what is the idea, haven't you heard that Emil Fischer proved that p are ppch?? Well Erice, he came complete with wife who oozed success from every pore: we had an exquisite din. as per usual, I engaged the wife in conversation until she sickened me w. every, yes every, damn thing in her life being just too-too (Ah yes? you know L Pauling, we of course spent a year there and Linderstrom Lang? ah yes we did a year with him in Copenhagen and what about Francis Crick? my husband sends his assistants there for a year or two' — It is really like a sort of Hollywood success story — and all these people back each other up at every turn, it being understood that this is price of membership in this success club. Well anyway, he grasped me by the arm, put his arm cosily round my waist and said 'Ah yes, how LONG have I known your name and now at last I meet you.' Well he gave his formal lec. and then, as usual there was to be a party for everyone + a discussion. Now this was too much to ask of me, I felt — and yet isn't it the devil that this should be so? But how could I talk in discussion and not put some of the qu.

*not sent*

regarding evidence for his views etc. . . . . this situation is repeated so unbelievably often that it makes discussion of what matters to me a sheer impossibility. You were writing a few weeks ago about my having no one to talk to at Smith — but this is only part of it in that I would be no better off at Princeton etc. as you seemed to think. They are, almost to a man, or all, without qualification, in this same state of mind. The prob. is CLOSED and they begin where the dogmas leave off and how can one start up things with any one of them, particularly in the atmosphere we have here on these gala occasions, so different from the dour cold atmosphere at the LMS and the RAS etc. Well all I try to do on these occasions is to prevent myself from engaging in this and that which will be entirely unprofitable. But these occasions 'kill me' as they say here and it takes me 24 hours to get back to where I was. Oh dear.

Erice in all these long years when no word regarding the things I care about have been exchanged between us I have been looking forward to the time which is now here and when the road is clear regarding time, money, etc. to get the ProtMon done — and now I can't do it. Isn't it hell.

Yes dearest Erice, will try to find Poirot on Holiday and am still trying about the other book you wanted.

Erice how does one get back self-confidence? This beats me. Yet I shall for ever remain unsatisfied if I can't get this ProtMon done. And this is really my contribution and is on such untrodden paths that it really should be on record. Why can't I get on with it? I never told you what a masterly job your analysis of a somewhat similar problem relating to someone else was, in my opinion, this being rather a hurting subject. But I would love, just love, to see the corresponding analysis in my case: I really would love you extra for it, and take it without flinching. I believe that it would help me very much and that I might get on better with ProtMon if you did this for me. So oh dear . . . Thank you about proofs but no WORD yet about when they will come etc. As ever regarding evidence for his views etc? So I just didn't go. Actually Erice I would be worse off at bigger place (more occasions like this one), except in so far as I could perhaps go to seminars etc but only on OTHER subjects. I expect you don't realize the extent to which my protein views are beyond the pale — except to the NSciF. at least to the extent of helping me to understand the small (very small) peptides. It is my belief if I can NOW write ProtMon that things would improve. I believe that they don't know what on earth I am talking about. I do so so much want to get things put down clearly and well and soon. So Erice dear any help will be a blessing to me, because I really can't face NOT being able to do this now that I have the time and money . . . Erice: you did a marvellous job in analysing how the self confidence business stands with a certain other person. Should have said this long ago but it is a hurting topic — well IF you could really just put down such an anal. in my case, I could learn from it I am sure and take it without flinching. Can you do this? Oh dear I just have to get this done and this would help me greatly. I do begin to understand your remarks about the operetta etc. oh dear. But I need more enlightenment. It would be such a boon There is no one to talk to about such a matter, just as there is no one to talk to here or elsewhere about cyclols — but as regards other things and indeed other parts of chem and biochem and pharm etc. here is very fair and WH is splendid. So dearest Erice don't worry about that side of things for me . . . Actually Erice the ProtMon is far more my contribution than ANYthing in MonOne. I wonder how much you know about ins and outs of the mineralogical side of ProtMon and about the theory of twinning etc which I devised to get the hemoglobin twins into a meaningful analysis etc. well anyway all the ProtMon stuff is on such untrodden paths that I MUST get it down. Maybe I shall be able to Erice, if you will encourage me a little as well as tell me sad things I ought to know about operatta talk etc.

Thank you so so much about proofs-to-be-but still no word from Munks. As ever

# CHRONOLOGY OF THE LIFE OF DOROTHY WRINCH

| | |
|---|---|
| **1894** | Born in Rosario, Argentina |
| **1913-17** | Studies mathematics and philosophy at Girton College, Cambridge, England (B.A., M.A. degrees) |
| **1918-20** | Lecturer in mathematics at University College, London (M.Sc. 1920, D.Sc. 1921) |
| **1920-24** | Yarrow Scientific Research Fellow, Girton |
| **1923-31** | Lecturer in mathematics, Lady Margaret Hall, Oxford (M.A. 1924, D.Sc. 1929) |
| **1922** | Marries John W. Nicholson, professor of physics at Oxford |
| **1928** | Daughter Pamela is born |
| **1931-33** | Studies geometry in Vienna, and physics, chemistry, and biology at various laboratories in Europe |
| **1935-40** | Rockefeller Foundation Research Fellow |
| **1937** | Marriage is dissolved |
| **1939** | Emigrates, with Pamela, to the United States |
| **1940-41** | Dept. of Chemistry, Johns Hopkins University |
| **1942** | Marries O.C. Glaser, Professor of Biology at Amherst |
| **1942-43** | First jointly appointed visiting professor in the Valley (at Smith, Amherst, and Mt. Holyoke Colleges) |
| **1943-65** | Special research professor of physics at Smith College |
| **1952** | O.C. Glaser dies |
| **1966-71** | Sophia Smith Fellow at Smith College |
| **1971** | Retires to Woods Hole |
| **1975** | Pamela Wrinch Schenkman dies in fire |
| **1976** | Dies at age 82 |

Dorothy Wrinch

Irving Langmuir, Dorothy Wrinch, Kathryn Blodgett. About 1938.

# SUGGESTED READING

## Symmetry

*On Growth and Form*, D'Arcy W. Thompson. Cambridge, England: Cambridge University Press, 1942; paperback edition (abridged), 1969.

*Lectures on the Principle of Symmetry*, Francis Jaeger. Amsterdam: Elsevier, 1920.

*Space Structures, Their Harmony and Counterpoint*, Arthur L. Loeb. Reading, Mass.: Addison-Wesley, 1976.

*Fantasy and Symmetry: The Periodic Drawings of M.C. Escher*, Caroline MacGillavry. New York: Abrams, 1976.

*Patterns of Symmetry*, Marjorie Senechal and George Fleck, eds. Amherst: University of Massachusetts Press, 1977. (Dedicated to the memory of Dorothy Wrinch.)

*Symmetry*, Hermann Weyl. Princeton: Princeton University Press, 1952.

## Structures of Matter

*Crystals: Their Role in Nature and in Science*, Charles Bunn. New York: Academic Press, 1966.

"The Three-Dimensional Structures of a Protein Molecule," John C. Kendrew, *Scientific American*, November 1961.

"The Hemoglobin Molecule," Max F. Perutz, *Scientific American*, November 1964.

"The Three-Dimensional Structure of an Enzyme Molecule," D.C. Phillips, *Scientific American*, November 1966.

*The Thread of Life*, John C. Kendrew. Cambridge, Mass.: Harvard University Press, 1966.

*The Architecture of Molecules*, Linus Pauling and Roger Hayward. San Francisco: W.H. Freeman, 1964.

"Fitness in the Universe: Choices and Necessities," George Wald, *Origins of Life*, **5,** 7 (1974).

*Order and Life*, Joseph Needham. New Haven: Yale University Press, 1936. Paperback edition, M.I.T. Press.

## Patterns in Science

"How Molecular Biology Started," John C. Kendrew, book review, *Scientific American*, March 1967, p. 141.

*The Social Construction of Reality*, Peter L. Berger and Thomas Luckmann. Garden City, N.Y.: Doubleday, 1966.

*The Double Helix: A Personal Account of the Discovery of the Structure of DNA*, James D. Watson. New York: Atheneum, 1968.

*Rosalind Franklin and DNA*, Anne Sayre. New York: Norton, 1975.

"The Growth of Science in Society," Michael Polanyi, *Minerva*, **5,** 533-545 (1967). Reprinted in *Knowing and Being*, by Michael Polanyi, Marjorie Green, ed. Chicago: University of Chicago Press, 1969.

"The Potential Theory of Adsorption," Michael Polanyi, *Science*, **141,** 1010-1013 (1963). Reprinted in *Knowing and Being*, op. cit.

"A Prophet Without Honor: Dorothy Wrinch, Scientist, 1894-1976," Marjorie Senechal, *Smith Alumnae Quarterly*, April 1977.

A discussion at the Symposium. Left to right: George Fleck, Mary Margaret Murphy, David Harker, two Smith students.
Back row: Martin Buerger, Caroline MacGillavry, Elsa Sichel.

Mary-Elizabeth Murdock and Claire Sullivan amidst the Wrinch Papers.

# NOTES

## INTRODUCTION

[1] Letter in support of an application for a Rhodes Traveling Fellowship, 1929. From the Dorothy Wrinch Collection.
[2] *Order and Life*, Joseph Needham. New Haven: Yale University Press, 1936. Paperback edition, M.I.T. Press.
[3] Letter, Carolyn Cohen to Marjorie Senechal, February 1977.
[4] Recorded at the Wrinch Symposium for the Woods Hole Historical Society Oral History Project.
[5] Letter, Dorothy Wrinch to Eric Neville, October 1940.

## SCULPTURAL MODELS, MODULAR SCULPTURES

[1] A.L. Loeb, The architecture of crystals, in "Module, Proportion, Symmetry, Rhythm" ed. G. Kepes, Braziller, N.Y. (1966).
[2] A.L. Loeb, Structure and Patterns in Science and Art, Leonardo 4, 339-346, 1971.
[3] E. Haughton and A.L. Loeb, "Symmetry: The case history of a program," *J. Res. in Science Teaching,* 1964, 132-145.
[4] A.L. Loeb and E. Haughton, "The programmed use of physical models," *J. Progr. Instruction, III,* 9-18.
[5] A.L. Loeb, "A Studies for Spacial Order," Proc. International Conference on Descriptive Geometry and Engineering Design Graphics, Fiftieth Anniversary Symposium of the Engineering Design Graphics Division of the ASEE, pp. 13-20, 1979.

## DECIPHERING PROTEIN DESIGNS

[1] Huizinga, J. *Homo Ludens.* Beacon Press, Boston (1955).
[2] Sarton, G. *The Life of Science.* Henry Schuman, New York (1948).
[3] Harvey, W. *The Circulation of the Blood* (tr. K.J. Franklin). Dent, London (1968).
[4] Jacob, F. *La Logique du Vivant (The Logic of Life,* tr. B. Spillander). New York (1973).

[5]Edsall, J.T. *Proteins, Amino Acids and Peptides* (Cohn, E.J. and Edsall, J.T.). Hafner Press, New York (1945).

[6]Bernal, J.D. "Opening Remarks." In Wolstenholme, G.E.W. and O'Connor, M. (eds.) *Principles of Biomolecular Organization* (Ciba Fdn. Symp.), 1-6. Churchill, London (1966).

[7]For many historical accounts, see *Fifty Years of X-ray Diffraction*, Ewald, P.P. (ed.) and numerous crystallographers. Published for the International Union of Crystallography by N.V.A. Oosthoek's Uitgeversmaatschappij, Utrecht, The Netherlands (1962).

[8]Astbury, W.T. "On the Structure of Biological Fibres and the Problem of Muscle." *Proc. Roy. Soc. B.* 134, 303-328 (1947).

[9]Bragg, W.L., Kendrew, J.C. and Perutz, M.F. "Polypeptide Chain Configuration in Crystalline Proteins." *Proc. Roy. Soc. A* 203, 321-357 (1950).

[10]Pauling, L., Corey, R.B. and Branson, H.R. "The Structure of Proteins: Two Hydrogen-Bonded Helical Configurations of the Polypeptide Chain." *Proc. Natl. Acad. Sci.* 37, 205-211 (1951).

[11]Bragg, W.H. *The Development of X-ray Analysis* (Phillips, D.A. and Lipson, H.F., eds.). Hafner Press, New York (1975).

[12]Cochran, W., Crick, F.H.C. and Vand, V. "The Structure of Synthetic Polypeptides. I. The Transform of Atoms on a Helix." *Acta Cryst.* 5, 581-586 (1952).

[13]Crick, F.H.C. "The Packing of $\alpha$-Helices: Simple Coiled-Coils." *Acta Cryst.* 6, 689-697 (1953).

[14]Cohen, C. and Holmes, K.C. "X-ray Diffraction for $\alpha$-Helical Coiled-Coils in Native Muscle." *J. Mol. Biol.* 6, 423-432 (1963).

[15]Pauling, L. and Corey, R.B. "The Pleated Sheet, a New Layer Configuration of Polypeptide Chains." *Proc. Natl. Acad. Sci.* 37, 251-285 (1951).

[16]Bernal, J.D. "My Time at the Royal Institution 1923-27." In Ewald, P.P. *op. cit.*, 522-525 (1962).

[17]Crowfoot-Hodgkin, D. and Riley, D.P. "Some Ancient History of Protein X-ray Analysis." In Rich, A. and Davidson, N. (eds.) *Structural Chemistry and Molecular Biology*, 15-28. W.H. Freeman & Co., San Francisco (1968).

[18]Rosenberg, J. *Rembrandt: Life and Work*. Phaidon Press, London (1964).

[19]Wrinch, D.M. "On the Pattern of Proteins." *Proc. Roy. Soc. A* 161, 507-546 (1937).

[20]Wrinch, D.M. "Fourier Transforms and Structure Factors." ASXRED Monograph #2. American Crystallographic Association (1946).

[21]Watson, J.D. *Molecular Biology of the Gene* (3rd ed.). W.A. Benjamin, Inc., California (1976).

[22]Bragg, W.L. "The Growing Power of X-ray Analysis." In Ewald, P.P., *op. cit.*, 120-135 (1962).

[23]Green, D.W., Ingram, V.M. and Perutz, M.F. "The Structure of Haemoglobin. IV. Sign Determination by the Isomorphous Replacement Method." *Proc. Roy. Soc. A* 225, 287-307 (1954).

[24] Kendrew, J.C., Bodo, G., Dintzis, H.H., Parrish, R.G., Wyckoff, H. and Phillips, D.C. "A Three-Dimensional Model of the Myoglobin Molecule Obtained by X-ray Analysis." *Nature* 181, 662-666 (1958).

[25] For a good discussion of the relation between science and contemporary art, including proteins as "large molecules organized into patterns which are extremely complex", see C.H. Waddington, *Behind Appearance: A Study of the Relations Between Painting and the Natural Sciences in this Century.* Edinburgh University Press, Edinburgh (1969).

[26] Perutz, M.F. "The Haemoglobin Molecule." *Scientific American* 211, #5, 64-76 (1964).

[27] Blake, C.C.F., Koenig, D.F., Mair, G.A., North, A.C.T., Phillips, D.C. and Sarma, V.R. "Structure of Hen Egg-White Lysozyme." *Nature* 206, 757-761 (1965).

[28] Richardson, J.S., Thomas, K.A., Rubin, B.H. and Richardson, D.C. "Crystal Structure of Bovine Cu,Zn Superoxide Dismutase at 3 Å Resolution: Chain Tracing and Metal Ligands." *Proc. Natl. Acad. Sci.* 72, #4, 1349-1353 (1975).

[29] Blow, D.M. "Structure and Mechanism of Chymotrypsin." *Accts. of Chem. Res.* 9, 145-152 (1976).

[30] Huber, R. and Bode, W. "Structural Basis of the Activation and Action of Trypsin." *Accts. of Chem. Res.* 11, 114-122 (1978).

[31] Monod, J., Wyman, J. and Changeux, J.P. "On the Nature of Allosteric Transitions: A Plausible Model." *J. Mol. Biol.* 12, 88-118 (1965).

[32] Baldwin, J.M. "Structure and Function of Haemoglobin." *Prog. Biophys. Molec. Biol.* 29, 225-320 (1975).

[33] Perutz, M.F. "Stereochemistry of Cooperative Effects in Haemoglobin." *Nature* 228, 726-734 (1970); "Structure and Mechanism of Haemoglobin." *Brit. Med. Bull.* 32, 195-208 (1976).

[34] Poljak, R.J., Amzel, L.M., Chen, B.L., Phizackerley, R.P. and Saul, F. "The Three-Dimensional Structure of the Fab' Fragment of a Human Myeloma Immunoglobulin at 2.0 Å Resolution." *Proc. Natl. Acad. Sci.* 71, #9 3440-3444 (1974); Poljak, R.J. "Three-Dimensional Structure, Function and Genetic Control of Immunoglobulins." *Nature* 256, 373-379 (1975).

[35] Schiffer, M., Girling, R.L., Ely, K.R. and Edmundson, A.B. "Structure of a $\lambda$-Type Bence-Jones Protein at 3.5 A Resolution." *Biochemistry* 12 4620-4631 (1973).

[36] Davies, D.R., Padlan, E.C. and Segal, D.M. "Three-Dimensional Structure of Immunoglobulins." *Ann. Rev. Biochem.* 44, 639-667 (1975).

[37] Colman, P.M., Dersenhofer, J., Huber, R. and Palm, W. "Structure of the Human Antibody Molecule Kol (Immunoglobulin G1): An Electron Density Map at 5 Å Resolution." *J. Mol. Biol.* 100, 257-282 (1976).

[38] Huber, R., Dersenhofer, J., Colman, P.M., Matsushima, Y. and Palm, W. "Crystallographic Structure Studies of an IgG Molecule and an Fc Fragment." *Nature* 264, 415-420 (1976).

[39] Ramachandran, G.N. "Stereochemistry of Biopolymer Conformation." In *Symmetry and Function of Biological Systems at the Macromolecular Level* (Nobel Symp. 11) (Engström, A. and Strandberg, B., eds.), 79-100. Wiley, New York (1968).

[40] Schultz, G.E., Barry, C.D., Friedman, J., Chou, P.Y., Fasman, G.D., Finkelstein, A.V., Lim, V.I., Ptitsyn, O.B., Kabat, E.A., Wu, I.T., Robson, B. and Nagano, K. "Comparison of Predicted and Experimentally Determined Secondary Structure of Adenyl Kinase." *Nature* 250, 140-142 (1974).

[41] Schultz, G.E. "Structural Rules for Globular Proteins." *Angewandte Chemie*, 16, #1, 23-32 (1977).

[42] Levitt, M. and Chothia, C. "Structural Patterns in Globular Proteins." *Nature* 261, 552-558 (1976).

[43] Jacob, F. "Evolution and Tinkering." *Science* 196, #4295, 1161-1166 (1977).

## THE STORY OF DNA

[1] J.D. Watson, *The Double Helix* (New York: Atheneum Publishers, 1968; Mentor Paperback, 1969).

[2] J.D. Watson and F.H.C. Crick, "A Structure for Deoxyribose Nucleic Acid," *Nature, 171*, 737-38, 1953.

[3] A. Sayre, *Rosalind Franklin and DNA* (New York: W.W. Norton and Co., 1975).

[4] R. Hubbard, "Rosalind Franklin and DNA" (Book Review), *Signs, 2*, 229-37, 1976.

[5] E.H. Gombrich, "Meditations on a Hobby Horse or the Roots of Artistic Form," in *Meditations on a Hobby Horse and Other Essays on the Theory of Art* (London: Phaidon Press, 1963).

[6] A.W. Watts, *Nature, Man and Woman* (New York: Pantheon Books, 1958; Vintage Books, 1970).

[7] Watson, paperback edition, p. 21.

[8] J.D. Watson, "Growing Up in the Phage Group," in *Phage and the Origins of Molecular Biology*, J. Cairns, G.S. Stent and J.D. Watson, eds. (Cold Spring Harbor, L.I.: Cold Spring Harbor Laboratory of Quantitative Biology, 1966).

[9] E. Schrödinger, *What is Life?* (London: Cambridge University Press, 1944).

[10] R. Olby, *The Path to the Double Helix* (London: Macmillan Co., 1974).

[11] F.H.C. Crick and J.D. Watson, "The Complementary Structure of Deoxyribonucleic Acid," *Proc. Roy. Soc., A 223*, 80-96, 1954.

[12] L. Pauling, "Fifty Years of Progress in Structural Chemistry and Molecular Biology," *Daedalus, 99*, 988-1014, 1970.

[13] R. Hubbard, "Preface to the English translations of Boll's *On the Anatomy and Physiology of the Retina* and of Kühne's *Chemical Processes in the Retina*," *Vision Research*, 17, 1247-48, 1977.

[14] F.H.C. Crick, *Of Molecules and Men* (Seattle and London, 1966, pp. 10, 14), as quoted in Olby, p. 425.

[15] M. Polanyi, "Life's Irreducible Structure," *Science*, 160, 1308-12, 1968, as quoted in Olby, p. 425.

[16] E. Chargaff, "Building the Tower of Babble," *Nature*, 248, 776-9, 1974.

## *ACKNOWLEDGMENTS*

The cover design for this book is a drawing of a fragment of the protein fabric postulated by Dorothy Wrinch as part of her cyclol theory.

Special thanks are due to the secretaries of the Clark Science Center at Smith College for their expert typing, and to my colleagues George Fleck and the other members of the Wrinch Symposium Planning Group, for helping to bring the symposium into being.

I am grateful to Dr. Mary-Elizabeth Murdock for permission to reproduce letters and photographs from the Wrinch Collection.

Claire Sullivan, Smith College class of 1978, did an excellent job of cataloguing the Wrinch Papers for the Sophia Smith Collection.

Publication of this book was made possible in part by a grant from the Friendship Foundation.

## DATE DUE

| | | | |
|---|---|---|---|
| | | | |
| | | | |
| | | | |
| SEP 6 1994 | | | |
| RETURNED | SEP 0 3 1994 | | |
| | | | |
| | | | |
| | | | |
| | | | |
| | | | |
| | | | |
| | | | |
| | | | |
| | | | |

DEMCO 38-297